Dangerous Science: Science Policy and Risk Analysis for Scientists and Engineers

Daniel J. Rozell

]u[

ubiquity press
London

Published by
Ubiquity Press Ltd.
Unit 322-323
Whitechapel Technology Centre
75 Whitechapel Road
London E1 1DU
www.ubiquitypress.com

First published 2020

Cover design by Amber MacKay
This cover has been designed using resources from Freepik.com
Cover illustration by Harryarts

Print and digital versions typeset by Siliconchips Services Ltd.

ISBN (Paperback): 978-1-911529-88-0
ISBN (PDF): 978-1-911529-89-7
ISBN (EPUB): 978-1-911529-90-3
ISBN (Mobi): 978-1-911529-91-0

DOI: https://doi.org/10.5334/bci

The full text of this book has been peer-reviewed to ensure high academic
standards. For full review policies, see http://www.ubiquitypress.com/

Suggested citation:
Rozell, D. J. 2020. *Dangerous Science: Science Policy and Risk Analysis
for Scientists and Engineers*. London: Ubiquity Press. DOI: https://doi.
org/10.5334/bci. License: CC-BY 4.0

To read the free, open access version of this
book online, visit https://doi.org/10.5334/bci
or scan this QR code with your mobile device:

Contents

Preface   ix

Case Study: H5N1 Influenza Research Debate   1

Assessing the Benefits of Research   9

Values in Risk Assessment   29

Technological Risk Attitudes in Science Policy   57

Managing Dangerous Science   77

Afterword   103

References   107

Index   141

Acknowledgments

Writing a book requires a substantial time commitment, and I would like to thank my family for their patience. It also requires resources, and I would like to thank Stony Brook University and the Department of Technology and Society in particular for its support. Thanks also goes out to David Tonjes for his helpful advice and Scott Ferson for his gracious support and instructive discussions regarding the nature of risk and the measurement of uncertainty. I must also express my deep gratitude to the late David Ferguson, instrumental in my academic career, who imparted kindness and wisdom in equal measure. Finally, I would like to thank Sheldon Reaven for our numerous conversations on the philosophy of science, science policy, and opera. The value of his insightful comments has been exceeded only by the value of his friendship.

This book has also benefitted from the insightful peer review comments provided by Professor Frederic Emmanuel Bouder.

Competing interests

The author declares that they have no competing interests in publishing this book.

Preface

It is hard to overstate the importance of science and technology in modern society. Humans have been reshaping their lives and environment with technology since the dawn of the species. The first stone tools and controlled use of fire even appear to predate *Homo sapiens*. There are countless technological touchstones along the path of human history that have fundamentally changed how we live: agriculture, metallurgy, alphabets, and so on. However, the rate of technological change has substantially increased in recent centuries. At the dawn of the Industrial Revolution, the steam engine, with its many impacts from railroads to factories, was at the forefront of social change. Subsequent technological advances continued to transform society. Imagine a world today without refrigeration, electric power, vaccinations, airplanes, plastics, or computers. In the early 21st century, the work of Silicon Valley was perhaps the most public image of technology. However, technology encompasses a much broader range of tools and techniques that humans employ to achieve goals.

Given this fundamental role of science and technology in our lives, relatively little public discussion is focused on science and technology policy. Politicians frequently declare support for science research and technology development but have much less to say regarding exactly how or what innovation should be encouraged or avoided.

It is equally surprising how little of science and engineering training is spent on the social impacts of research and development. Most engineering students in the US are required to take a token course in engineering ethics, often a business ethics course tailored to engineers, which is treated separately from the technical coursework. Meanwhile, the physical sciences frequently require no training at all, while biological and social sciences usually require a short course in the appropriate treatment of animal or human test subjects. Some federal research grants require training in the responsible conduct of research—often weakly implemented (Phillips et al. 2018)—which focuses on the research and publication process. What happens after publication is given only cursory attention.

Few people would argue that scientists and engineers bear absolutely no responsibility for how their work is used. Yet the potential use of research can be difficult to predict, so it is also hard to argue that they bear total responsibility. So what is their level of culpability and what should they do? As with other tough questions without clear answers, the typical result is to politely ignore the issue or label it someone else's jurisdiction. However, inaction comes with a price. Scientists and engineers, well-trained and comfortable in the lab or field, will occasionally find themselves under public scrutiny with inadequate training in science policy and risk analysis.

A particularly stressful scenario is when researchers announce or publish work only to receive a decidedly negative public reaction. No one wants their dedicated efforts, in which they take great pride, to be seen as dangerous science, but it happens. Research can be viewed as dangerous in either its practice or in its results. This danger can be broad. Some technologies may pose physical danger to humans or the environment. Other technologies are morally dangerous—they violate a common societal value, make crossing an ethical line easier, or simply cause more harm than benefits.

I hesitated to call this book 'Dangerous Science' because I did not want to alienate the intended audience with an alarmist and vaguely anti-science-sounding title. However, terms such as 'controversial research' or 'unpopular technology' do not quite capture the full impact of science and technology that meets public opposition. This book is intended for scientists and engineers, and it is important for this audience to understand that the science and technology that they work on could be dangerous on many levels. For example, real or perceived dangers to society can translate into real dangers to the careers of individual scientists and engineers.

The book's subtitle is just as important. This is an introductory, but not simplistic, guide to science policy and risk analysis for working scientists and engineers. This book is not for experts in science policy and risk analysis: it is for the biologist considering synthetic biology research; it is for the computer scientist considering autonomous military equipment development; it is for the engineers and atmospheric scientists considering geoengineering responses to climate change. The public has already found some of the research within these fields objectionable, and it is wise to enter the policy arena prepared.

Scientists and engineers must be cognizant of cultural sensitivities, ethical dilemmas, and the natural potential for unwarranted overconfidence in our ability to anticipate unintended harm. This task can be both simple, yet difficult. On one hand, the writings and conversations of many scientists are often deeply self-reflective and nuanced. Yet, the fundamental core of science and engineering is empirical, analytical, and often reductionist—traits that can work against making connections between technology and society.

Science is often used to make public policy, and there is a perennial effort to increase 'evidence-based' public policy (Cairney 2016). Although they are related tasks, this book does not focus on how to use science to make good policy but rather how to use policy to make good science. Specifically, it explores the idea of dangerous science—research that faces public opposition because of real or perceived harm to society—and why debates over controversial research and technology are not easily resolved. More importantly, it also suggests techniques for avoiding a political impasse in science and technology policymaking. The target audience of this book is future or working scientists and engineers—people who care deeply about the impact of their work but without the time to fully explore the fields of science policy or risk analysis (it is hard enough being an expert in one field).

The French polymath Blaise Pascal was one of many authors to have noted that they would have preferred to write more briefly if only they had the time to shorten their work. Given that the seeds of this book were formed about a decade ago, I've made considerable effort to distill this work down to something not overly imposing to the reader while avoiding the oversimplification of complex issues. The intent here is not to drown the reader in minutiae, but rather to lead the reader through an overview of the difficulties of assessing and managing science and technology with ample references for further reading as desired. The hope of such brevity is that it will actually be read by the many busy professionals who need to consider the societal impact of potentially dangerous science and do not want to find themselves unprepared in the middle of a political maelstrom.

While a primary audience of the book is the graduate student looking for a supplement to standard 'responsible conduct of research' required reading (e.g., Institute of Medicine 2009), the book is also written to be approachable by anyone interested in science policy. If there is one overarching lesson to be taken from this book, it is that science in the public interest demands public involvement. Science and technology have become too powerful to engage in simple trial and error experimentation. Before action, thoughtful consideration is required, and this benefits from as many ideas and perspectives as possible. Oversight must now be a communal activity if it is to succeed.

The general form of the book is laid out in the following sequence.

In the first chapter, a case study is presented that walks through the events and fundamental issues in one dangerous science example. In this case, it was the debate, starting in 2011, over gain-of-function research involving the H5N1 avian influenza virus that sparked public fears of a potential accidental pandemic. The description of the multi-year debate demonstrates the practical difficulties of assessing and managing dangerous science. It ends with the question of why a formal risk-benefit analysis commissioned for the debate failed to resolve the controversy.

In the second chapter, we tackle one part of that question and review the many ways in which the benefits of research can be defined. In addition, comparing the various methods of estimating the benefits of research can provide insight into how science policy is formulated. We review data-driven methods of assessing research benefits, including estimating the effects of research on job production, economic growth, scientific publications, or patents. More subjective methods, such as value-of-information analysis and expert opinion, have also been recommended to account for less quantifiable benefits and public values. A comparison of the various legitimate, but essentially incomparable, ways that research benefits are assessed suggests that no form of assessment can be both quantitative and comprehensive. Discussing the strengths and weaknesses of each approach, I argue there is currently no reliable or universally acceptable way of valuing research. The result is that formal assessments of research benefits can be useful for informing public science policy debates but should not be used as science policy decision criteria.

In the third chapter, we tackle the other half of the risk-benefit debate by reviewing the many factors that can compromise the perceived legitimacy of a risk assessment. Formal risk assessment is often idealized as objective despite many warnings that subjective value judgments pervade the risk assessment process. However, prior warnings have tended to focus on specific value assumptions or risk assessment topics. This chapter provides a broad review of important value judgments that must be made (often unknowingly) by an analyst during a risk assessment. The review is organized by where the value judgments occur within the assessment process, creating a values road map in risk assessment. This overview can help risk analysts identify potentially controversial assumptions. It can also help risk assessment users clarify arguments and provide insight into the underlying fundamental debates. I argue that open acknowledgment of the value judgments made in any assessment increases its usefulness as a risk communication tool.

In the fourth chapter, we acknowledge that policy formulation for controversial science and technology must often occur in the absence of convincing evidence. As a result, technology policy debates frequently rely on existing technological risk attitudes. I roughly categorize these attitudes as either technological optimism or skepticism and review multiple theories that have been proposed to explain the origins of these attitudes. Although no individual theory seems to provide a complete explanation so far, we do know that technological risk attitudes are flexible and influenced by a complex range of factors that include culture and personal circumstances. An important result of these opposing attitudes is that moral arguments against dangerous science are often downplayed and policymakers tend to act cautiously permissive. Several emerging technologies, such as human genome editing, synthetic biology, and autonomous weapons, are discussed in the context of technological risk attitudes.

In the fifth and last chapter, we turn to potential solutions for managing science controversies. After briefly reviewing traditional risk management techniques, I argue dangerous science debates should place less emphasis on attempting to quantify risks and benefits for use as a decision tool. Rather risk-benefit assessments are better used as risk exploration tools to guide better research design. This is accomplished by engaging multiple perspectives and shifting away from traditional safety and security measures toward more inherently safe research techniques that accomplish the same goals. The application of these principles are discussed in the example of gene drive technology.

One final remark on the book's contents: There is plenty of ammunition in this book for science 'denialists' if they engage in cherry-picking. Despite the critiques of particular lines of research or methods presented here, this book is not an attack on the enterprise of science, which has incalculable practical and intellectual value to society. Generally, more science is better. However, it is antithetical to the progress of science to take the authoritarian approach of 'You're either with us or against us' and to avoid all valid criticisms of how science is conducted. The purpose here is to improve the process for assessing and managing the broader impacts of science. Open discussion is the only way forward.

Case Study: H5N1 Influenza Research Debate

Media coverage of the latest scientific discoveries and technological innovations is usually enthusiastic. Long-standing questions are answered, productivity is increased, health is improved, and our standard of living is raised—all due to modern science and human ingenuity. However, sometimes the results of research can also inspire public fear and outrage. Let us consider a recent example.

In September 2011, Dutch virologist Ron Fouchier announced at a conference in Malta that his research team had recently engineered a version of the H5N1[1] avian influenza virus that was highly transmissible between mammals. Shortly thereafter, virologist Yoshihiro Kawaoka of the University of Wisconsin presented results from a similar study. The subsequent media coverage scared the general public and set off demands for a review of how science research is assessed, funded, and managed.

Why would these studies be so scary? The reason centers on the lethal potential of influenza and the H5N1 virus in particular. An influenza pandemic typically occurs when a flu virus substantially different from circulating viruses mutates to become easily transmissible between humans. The combination of

---

[1] The formal naming convention for influenza viruses includes the antigenic type (A, B, or C); the originating host (e.g., swine); the geographical origin; year of identification; strain number; or for type A viruses, the specific subtypes of two surface proteins, hemagglutinin (H) and neuraminidase (N) (e.g., H5N1) (Assaad et al. 1980). Given the complexity of formal names, popular shorthand names, such as Spanish flu, Swine flu, and H1N1 can potentially be referring to the same influenza virus. The World Health Organization has been working to improve shorthand names to make them less stigmatizing and more informative.

How to cite this book chapter:
Rozell, D. J. 2020. *Dangerous Science: Science Policy and Risk Analysis for Scientists and Engineers*. Pp. 1–8. London: Ubiquity Press. DOI: https://doi.org/10.5334/bci.a. License: CC-BY 4.0

limited natural immune response and quick transmission is dangerous. The 1957 and 1968 influenza pandemics killed about 0.1 percent of the individuals infected, which resulted in over one million fatalities worldwide in each case. The relatively mild 2009 influenza pandemic's mortality rate was half that. The infamous 1918 influenza pandemic, which killed tens of millions worldwide, had an estimated lethality of 2 to 5 percent. Although highly uncertain, estimates of H5N1 lethality range from 1 to 60 percent (Li et al. 2008; Wang, Parides & Palese 2012). The few cases of H5N1 influenza in humans have been primarily attributed to direct contact with birds that harbored the virus. If the virus gained the ability to easily transmit between humans while retaining its lethality, the results would be catastrophic. This is exactly what the H5N1 studies appeared to create and why the public reaction was so negative. Although there are many pathogens capable of creating pandemics, the influenza virus is exceptional in that it tends to be easily transmissible and cause high rates of infection with a virulence that ranges from mild to deadly. For these reasons, a major influenza pandemic has been placed on the short list of truly catastrophic global events that includes nuclear war, climate change, and large meteor strikes (Osterholm & Olshaker 2017; Schoch-Spana et al. 2017).

The response to the H5N1 research announcements shows the difficulty of assessing and managing potentially dangerous research. The purpose of the two bird flu studies (Herfst et al. 2012; Imai et al. 2012), both funded by the US National Institutes of Health (NIH), was to investigate how easily the H5N1 virus could naturally become a more serious public health threat. However, many scientists and security experts became concerned that the papers detailing the experiments could be a blueprint for skilled bioterrorists. In November 2011, the US National Science Advisory Board for Biosecurity (NSABB) recommended redacting the methodology for each paper. This was the first recommendation of publication restriction since the board's formation in 2005. The following month, the Dutch government, based on a law—Council Regulation (EC) No 428/2009—aimed at weapons nonproliferation, requested that Dr. Fouchier apply for an export license before publishing his research. Although the license was granted within days, the unprecedented application of the law to virology research was shocking to many in the science community. As a result, a voluntary H5N1 research moratorium was agreed upon by prominent influenza research laboratories in January 2012 until new guidelines could be put in place. The following month, a review by a panel of experts at the World Health Organization (WHO) contradicted the NSABB by recommending full publication of the research. The NSABB performed a second review in March 2012 and reversed its position by also recommending full publication. Critics viewed this sudden change as acquiescence to pressure from the scientific community.[2]

---

[2] An insider's perspective of the NSABB decision process and the pressures it faced can be found in *Deadliest Enemy* (Osterholm & Olshaker 2017).

Along with the NSABB reversal, the NIH simultaneously released a new policy, *US Government Policy for Oversight of Life Sciences Dual Use Research of Concern*, as guidance for institutional biosafety committees. The four-page document clarified what research counted as 'dual-use research of concern'— that is, research that has both potential societal benefits and obvious malicious uses. Almost a year later, new guidelines were released by the US Department of Health and Human Services (HHS) in February 2013 for funding H5N1 gain-of-function research. In this case, gain-of-function means the purposeful mutation of a disease agent to add new functions or to amplify existing undesirable functions, such as transmissibility or virulence. Among the various new requirements, the HHS policy mandated the following: the research address an important public health concern; no safer alternative could accomplish the same goal; biosafety and biosecurity issues were addressed; and risk reduction oversight mechanisms be put in place (Patterson et al. 2013). The review process was extended to H7N9 bird flu research in June 2013.

While these clarifications would appear to have settled the matter, they did not. One reason is, like the various policies before it, the HHS policy did not address exactly *how* to assess the risks and benefits of a research proposal. Back in 2006, the NSABB made a similar move when it asked authors, institutional biosafety committees, and journal editors to perform a risk-benefit analysis before publishing dual-use research, without providing detailed guidance on how to perform such an analysis. Five years later, an NIH random survey of 155 life science journals found that less than 10 percent had a written dual-use policy or reported reviewing dual-use manuscripts in the previous 5 years (Resnik, Barner & Dinse 2011). Likewise, a 3-year sample of 74,000 biological research manuscripts submitted to Nature Publishing Group resulted in only 28 flagged and no rejected manuscripts for biosecurity concerns (Boulton 2012).

While it is possible that dual-use research is quite rare, it is more likely the research community is just unable to recognize research as dangerous due to lack of training and proper guidance (Casadevall et al. 2015). In light of publications, such as the papers detailing the synthesis of poliovirus (Cello, Paul & Wimmer 2002) and the reconstruction of the 1918 flu virus (Tumpey et al. 2005), perhaps the H5N1 influenza papers are most notable in that they actually started a significant public debate over biosecurity policy (Rappert 2014).

Thus, it is not surprising that the new HHS policy did not resolve the controversy, and three papers published in 2014 renewed the debate surrounding gain-of-function flu research. One study made the H7N1 flu strain, which was not covered by the new HHS rules, transmissible in mammals (Sutton et al. 2014). A second study by Dr. Fouchier's lab expanded on earlier H5N1 work (Linster et al. 2014). The third paper, published by Dr. Kawaoka's lab, detailed the engineering of a virus similar to the strain responsible for the 1918 flu pandemic to argue that another major pandemic could arise from the existing reservoir of wild avian flu viruses (Watanabe et al., 2014). Many critics were particularly disturbed by this last paper because the University of

Wisconsin biosafety review of the proposal failed to classify the work as 'dual-use research of concern' despite a consensus among biosecurity experts that it clearly was. Collectively, these results again raised concerns that an intentional or accidental release of an engineered virus from gain-of-function influenza research could be the source of a future pandemic—an ironic and deadly self-fulfilling prophesy.

The Emphasis Shifts from Terrorists to Accidents

Events in 2014 brought the H5N1 debate back to the popular media but shifted the primary concern from biosecurity to biosafety. First, an accidental exposure of multiple researchers to anthrax bacteria was discovered at the US Centers for Disease Control and Prevention (CDC) in Atlanta. Shortly thereafter, it was announced there had been an accidental contamination of a weak flu sample with a dangerous flu strain at another CDC lab that further jeopardized lab workers. This led to the temporary closure of the CDC anthrax and influenza research labs in July 2014 and the resignation of the head of a bioterrorism lab.

The final straw may have been the discovery of six vials of live smallpox virus in storage at a US Food and Drug Administration (FDA) lab in Bethesda, Maryland, in June 2014. After smallpox was globally eradicated in 1980, only two secure labs (the CDC in Atlanta, Georgia and the Vector Institute in Novosibirsk, Russia) were supposed to have smallpox virus samples. The planned destruction of these official samples had been regularly debated by the World Health Assembly for decades (Henderson & Arita 2014). Finding unsecured samples elsewhere was rather disturbing.

After these events, it became clear that human error was much more prevalent at the world's top research facilities than previously believed. This prompted CDC director Thomas Frieden to suggest closing many biosafety level 3 and 4 labs.[3] Unfortunately, there was no definitive list of these labs for public review or government oversight (Young & Penzenstadler 2015). However, it was estimated that the number of labs working with potentially pandemic pathogens had tripled in this century. This resulted in the number of reported lab accidents and the number of workers with access or exposure risk to increase by at least an order of magnitude. According to a 2013 US Government Accountability Office assessment, the increased number of labs and lab workers had unintentionally increased rather than decreased national risk.

These mishaps led the US Office of Science and Technology Policy and the HHS to impose yet another temporary moratorium on NIH-funded gain-of-function research for influenza in October 2014. The moratorium was

---

[3] Biosafety levels (BSL) range from BSL-1 to BSL-4, with the latter having the strictest protocols and equipment for handling potentially fatal infectious agents for which there are no vaccines or treatment.

intended to last until the NSABB and National Research Council (NRC) could assess the risks and benefits of these lines of research.

The Risk-Benefit Assessment

The moratorium called for 'a robust and broad deliberative process' to 'evaluate the risks and potential benefits of gain-of-function research with potential pandemic pathogens' (OSTP 2014). The proposed multi-step process consisted of a series of meetings by the NSABB and NRC. The meetings would first draft recommendations on how to conduct a risk-benefit analysis and then evaluate an independently performed risk-benefit assessment and ethical analysis (Selgelid 2016). The primary role of the NRC was to provide additional feedback from the scientific community while the NSABB provided the final recommendation to the Secretary of HHS and director of the NIH.[4] A $1.1 million contract was awarded to an independent[5] Maryland biodefense consulting firm, Gryphon Scientific, in March 2015 to conduct the risk-benefit analysis. The assessment was to be 'comprehensive, sound, and credible' and to use 'established, accepted methods in the field' (NIH 2014). The draft assessment (Gryphon Scientific 2015), completed in eight months, was over 1,000 pages long. It included a separate quantitative biosafety risk analysis, a semi-quantitative biosecurity risk analysis, and a qualitative benefits assessment. The main finding of the report was the majority of gain-of-function research posed no more risk than existing wild-type influenza strains. However, for a strain as pathogenic as the 1918 influenza, the biosafety risk analysis estimated that an accidental laboratory release would result in a global pandemic every 560 to 13,000 years resulting in up to 80 million deaths. The biosecurity assessment was harder to quantify but estimated a similar risk if the theft of infectious material by malevolent actors occurred at least every 50 to 200 years.

An NRC symposium (NRC 2016) was convened in March 2016 to further discuss the risk-benefit assessment and the draft recommendations proposed by the NSABB. Comments made during the symposium, as well as public comments received directly by the NSABB, were generally critical of the Gryphon report. Criticisms included claims that both the completion of the report and the review process were rushed, the report ignored existing data, the report missed alternative methods of infection, and it communicated results in units that obscured risk. Some members of the scientific community believed the

---

[4] Yes, that is a lot of acronyms.
[5] An independent third-party analysis should not be confused with an impartial analysis. Analysts can be influenced by the hiring organization, social pressures, and cultural norms. Furthermore, analysts trained and embedded in the field of interest are more likely to have personal positions at the outset—it is not easy to be a truly impartial expert.

report was comprehensive and balanced because parties on both sides of the debate were not fully satisfied with the report (Imperiale & Casadevall 2016). However, unlike policy debates, the final goal of science is truth-seeking, not compromise. Thus, there seemed to be some confusion regarding whether the risk-benefit assessment was a technical or policy tool.

The final NSABB report (NSABB 2016) contained a series of findings and recommendations. The report found that most types of gain-of-function research were not controversial and the US government already had many overlapping policies in place for managing most life science research risks. However, not all research of concern was covered by existing policies. More importantly, there was some research that should not be conducted because the ethical or public safety risk outweighed the benefits. Unfortunately, no specific rule of what constitutes unacceptable research could be formulated. Rather, the report stated the need for 'an assessment of the potential risks and anticipated benefits associated with the individual experiment in question.' This essentially meant that any study in question would require its own *ad hoc* risk-benefit assessment process. The NSABB report provided a short list of general types of experiments that would be 'gain-of-function research of concern' requiring additional review, as well as types of experiments that would not, such as gain-of-function research intended to improve vaccine production. While this provided more clarification than what was available at the start of the process, public comments from the NRC and NSABB meetings tended to show a general dissatisfaction with the lack of specificity and vagueness of terms, such as 'highly transmissible.' In particular, the critics were hoping for a detailed list of types of dangerous experiments that should not be conducted. There was also concern that the definition of 'gain-of-function research of concern' was too narrow. Based on the new criteria, it appeared that the original research that started the debate—creating a mammalian-transmissible avian influenza virus—would not be considered research of concern. Finally, there was also fear that assessments would not be conducted by disinterested third parties with appropriate expertise.

In the end, the competing stakeholders had not reached consensus: the proponents saw the additional regulatory processes as unnecessary, while the critics saw the new recommendations as grossly insufficient. The fundamental problem was that the competing experts still did not agree on even the basic utility or level of danger associated with the research. Consequently, it was not possible to create a defensible quantitative risk-benefit assessment or subsequent science research policy that was endorsed by most of the stakeholders.

In January 2017, the Office of Science and Technology Policy issued a policy guidance document (OSTP 2017) that lifted the gain-of-function research moratorium for any agency that updated review practices to follow the NSABB recommendations. The NIH officially announced it would once again consider funding gain-of-function research in December 2017. At the same time, HHS issued a six-page guidance document, *Framework for Guiding Funding*

Decisions about Proposed Research Involving Enhanced Potential Pandemic Pathogens (HHS P3CO Framework), which outlined the extra layer of consideration required for future funding decisions for applicable research. The US government approved the continuation of gain-of-function research at the labs of both Dr. Kawaoka and Dr. Fouchier in 2018 (Kaiser 2019).

So, over the course of six years, six public NSABB meetings, two public NRC meetings, a year-long independent formal risk-benefit analysis, and a three-year pause on NIH funding for specific gain-of-function viral research, the research community found itself essentially back where it started in terms of how to actually assess and manage dangerous research. Proponents considered the matter settled, while critics saw the regulatory exercise as a fig leaf for continuing business as usual. The only significant improvement was the influenza research community was now a bit wiser about the limitations of the science policy-making process.

One could argue that this is a rather harsh assessment of the substantial efforts of many individuals, but even if we take the optimistic view that the gain-of-function research debate refocused the attention of the scientific community on the importance of assessing and managing dangerous science research, then how do we explain continuing research controversies? For example, in early 2018, virologists from the University of Alberta published a paper describing the *de novo* synthesis of an extinct horsepox (Noyce, Lederman & Evans 2018). Claims that the work could lead to a safer smallpox vaccine were met with skepticism considering a safe vaccine already exists and there is no market for a new one (Koblentz 2018). Despite the paper passing a journal's dual-use research committee, critics argued that the paper lays the groundwork for recreating the smallpox virus from scratch (Koblentz 2017). If reactive *ad hoc* narrow regulations worked, then the scientific research community would not continue to be surprised by such seemingly reckless studies on a regular basis.

A Larger Question

This case study contains some valuable lessons regarding science policy, the responsible conduct of research, and the need to consider the implications and public perception of dangerous research. One obvious lesson is that science policy-making is a political decision-making process and not simply a matter of sufficient data and analysis. This was best summarized by risk expert Baruch Fischhoff, an NRC symposium planning committee member, when he said, 'Anybody who thinks that putting out a contract for a risk-benefit analysis will tell the country what to do on this topic is just deluding themselves' (NRC 2015). However, it is less obvious why this is so. After all, insurance companies have been conducting quantitative risk assessments for decades—why is this any different?

The US government response to the gain-of-function influenza research controversy was a typical response to dangerous science—an *ad hoc* directive to assess and monitor research in its early stages without details on how to proceed. Strangely, the lack of clear procedures for conducting a risk-benefit assessment extends to even relatively narrow and uncontroversial research questions that commonly come before regulatory agencies, such as medical drug efficacy comparisons (Holden 2003). This means that scientists and engineers are often performing incomparable assessments (Ernst & Resch 1996). Risk assessments clearly have value in so far as they focus attention on the public impact of research. However, it is not obvious that merely asking scientists to consider the risks and benefits of their work will result in due consideration and communication of risks in ways that satisfy policymakers and the general public (Fischhoff 1995). Risk-benefit analysis is often recommended as a policy mechanism for mitigating technological risk, but it is still unclear what its practical value is to policy formulation. This leads to the fundamental question—how *do* we assess the risks and benefits of potentially dangerous science?

Assessing the Benefits of Research

To answer the question of how we assess the risks and benefits of dangerous science, it helps to break down the problem. We will start with the assessment of benefits—a topic frequently revisited during budgetary debates over government funding of research. Trying to assess the benefits of research is a long-standing and contentious activity among science policy analysts and economists. Government-funded research constitutes less than one third of total research spending in the US, but public funding of research does not merely augment or even displace private investment (Czarnitzki & Lopes-Bento 2013). Rather, public funding is critical to early-stage, high-risk research that the private sector is unwilling to fund. The result is a disproportionate contribution of government funding to technological innovation (Mazzucato 2011). For example, while the private sector funds nearly all the pharmaceutical clinical trials in the US, public funding is still the largest source of pharmaceutical basic research.

Methods of Valuation

So how do policymakers assess the benefits of publicly funded research? Let's look at some of the most common approaches.

Research as a jobs program

In a simple input-output model of research, spending on salaries, equipment, and facilities associated with research has an analogous output of research jobs, manufacturing jobs, construction jobs, and so forth, but the impact of the actual research output is neglected (Lane 2009). While this is a simplistic way to

---

How to cite this book chapter:
Rozell, D. J. 2020. *Dangerous Science: Science Policy and Risk Analysis for Scientists and Engineers.* Pp. 9–27. London: Ubiquity Press. DOI: https://doi.org/10.5334/bci.b.
License: CC-BY 4.0

view research, it is popular for two reasons. First, it is relatively quantifiable and predictable compared to methods that focus on research output. For example, the STAR METRICS[1] program was started in 2009 to replace anecdotes with data that could be analyzed to inform the 'science of science policy' (Largent & Lane 2012). However, the first phase of STAR METRICS only attempted to measure job creation from federal spending (Lane & Bertuzzi 2011; Weinberg et al. 2014). A subsequent program, UMETRICS, designed to measure university research effects, used a similar approach by analyzing the same STAR METRICS data to determine the job placement and earnings of doctorate recipients (Zolas et al. 2015).

The second reason for the jobs-only approach is because job creation and retention is a primary focus of government policymakers. Elected officials may talk about the long-term implications of research spending, but the short-term impacts on jobs for their constituents are far more relevant to their bids for re-election. As US Representative George E. Brown Jr. noted, the unofficial science and technology funding policy of Congress is 'Anything close to my district or state is better than something farther away' (Brown 1999).

One outcome of viewing research in terms of immediate job creation is any research program may be seen as a benefit to society because all research creates jobs. However, this ignores that some research, through automation development or productivity improvements, will eventually eliminate jobs. Likewise, when research funding is focused on the desire to retain science and engineering jobs in a particular electoral district, it can diminish the perceived legitimacy of a research program. For example, there is a long-standing cynical perception of some US National Aeronautics and Space Administration (NASA) funding acting as a southern states jobs program (Berger 2013; Clark 2013).

Econometric valuation

The jobs-only perspective is obviously narrow, so most serious attempts at measuring the benefits of research use broader economic indicators. Econometric methods have attempted to measure the value of research by either a microeconomic or macroeconomic approach.

The microeconomic approach attempts to estimate the direct and indirect benefits of a particular innovation, often by historical case study. The case study approach offers a depth of insight about particular technologies that is often underappreciated (Flyvbjerg 2006). However, it is time and resource intensive, and its detailed qualitative nature does not lend itself to decontextualized

---

[1] Science and Technology for America's Reinvestment - Measuring the EffecTs of Research on Innovation, Competitiveness, and Science (a bit of a stretch for an acronym)

quantification.[2] Furthermore, actual benefit-cost ratios or rates of return for case studies tend to be valid only for the industry and the time period studied. As a result, they can be a poor source for forming generalizations about research activities. Additionally, innovation often comes from chance discovery (Ban 2006), which further complicates attempts to directly correlate specific research to economic productivity.

The macroeconomic approach attempts to relate past research investments to an economic indicator, such as gross domestic product (GDP).[3] This approach is more useful for evaluating a broader range of research activities. Using the macroeconomic approach, the value of research is the total output or productivity of an organization or economy based on past research investments. Three important factors have been noted when attempting a macroeconomic valuation of research (Griliches 1979):

1. The time lag between when research is conducted and when its results are used defines the timeframe of the analysis. Depending on the research, the time lag from investment to implementation may take years or decades.
2. The rate at which research becomes obsolete as it is replaced by newer technology and processes should be considered. The knowledge depreciation rate should be higher for a rapidly changing technology than for basic science research. For example, expertise in vacuum tubes became substantially less valuable after the invention of the transistor. Conversely, the value of a mathematical method commonly used in computer science might increase over time.
3. There is a spillover effect in research based on the amount of similar research being conducted by competing organizations that has an impact on the value of an organization's own research. This effect might be small for unique research that is unlikely to be used elsewhere. Further complicating this effect is the influence of an organization's 'absorptive capacity' or ability to make use of research output that was developed elsewhere (Cohen & Levinthal 1989). Even without performing substantial research on its own, by keeping at least a minimum level of research capability, an organization can reap the benefits of the publicly available research output in its field.

---

[2] This has not stopped big-data enthusiasts from trying. For example, keyword text-mining was performed on a 7,000 case study audit of research impacts in the United Kingdom. Ironically, the point of the audit was to add complementary context to a quantitative assessment (Van Noorden 2015).

[3] While GDP is a popular economic indicator, detractors dislike its simplicity which ignores many factors, including inequality, quality-of-life, happiness, and environmental health (Masood 2016; Graham, Laffan & Pinto 2018).

In general, quantifying any of the above factors is easier for applied research than for basic research. Likewise, it is easier to quantify private benefit to a particular organization than public benefit. Another factor that prevents easy identification of the economic value of research is the general lack of data variability. Research funding rarely changes abruptly over time, so it is difficult to measure the lag between research investments and results (Lach & Schankerman 1989).

The most common approach for determining the economic rate of return for research is growth accounting where research is assumed to produce all economic growth not accounted for by other inputs, such as labor and capital. Economists often refer to this unaccounted growth as the Solow residual (Solow 1957). A comprehensive review (Hall, Mairesse & Mohnen 2009) of 147 prior research studies that used either an individual business, an industry, a region, or a country to estimate the rate of return of research found a variety of results. The majority of studies found rates of return ranging from 0 to 50 percent, but a dozen studies showed rates over 100 percent[4]—a wide interval that portrays the difficulty of quantifying the benefits of research.

Not surprisingly, the return on research is not constant across fields, countries, or time, so any estimates from one study should be used cautiously elsewhere. Likewise, it is important to distinguish general technological progress from research. While technological progress may account for most of the Solow residual, a non-negligible amount of innovation occurs outside of funded research programs (Kranzberg 1967, 1968). Ultimately, due to the many potential confounding factors, such as broader economic conditions or political decisions, it is difficult to show a causal relationship for any correlation of productivity or profit with research.

Unfortunately, some advocacy groups have issued reports that imply a simple direct relationship between scientific investment and economic growth. Such statements are unsupported by historical data (Lane 2009). For example, Japan spends a higher proportion of GDP on research than most countries but has not experienced the expected commensurate economic growth for the past two decades. Likewise, at the beginning of this century, research spending in the US was about 10 times higher than in China. As a result, US contribution to the global scientific literature was also about 10 times higher than China's. However, in the following decade, China's economy expanded 10 times faster than the US economy.[5] The exact relationship between research spending and economic growth remains unclear; the only consensus is that research is beneficial.

---

[4] One outlier study conservatively (or humbly) estimated rates between -606 and 734 percent.

[5] Ironically, China's robust economy allowed it to dramatically increase its research spending and eventually surpass the US in total number of science publications according to NSF statistics.

Valuation by knowledge output

Given the difficulties of using econometric methods to assess research, economists have explored other methods that avoid monetizing research benefits
and the private versus social benefit distinction. One popular alternative
is to use academic publications. Despite its relative simplicity compared to
economic growth, publications are still problematic. First, comparisons are
complicated because scientific publication is not equally valued among all
fields and organizations. Publication in prestigious journals is often essential
to career advancement in academia but is relatively unimportant for industrial research scientists. Likewise, an ever-increasing proportion of research
is being disseminated outside standard academic journals via the internet—
open access pre-print archives, research data repositories, and code-sharing
sites have all become common. It is unclear how these new modes of information sharing should be measured. Second, the method does not assess the
relative value or visibility of an individual publication. This issue is partially
addressed by using the number of citations rather than the number of publications. However, citations are not a clear sign of quality research. For example, citations are commonly made to provide basic background information
(Werner 2015). Similarly, a journal's impact factor—the average number of
citations for a journal's articles in the past year—is widely derided as a proxy
for research output quality (yet the practice is still shamefully common). Conversely, a lack of citations does not necessarily indicate a lack of social benefit.
The information in an academic research article may be widely used in non-
academic publications—reports, maps, websites, and so forth—without ever
generating a citation that can be easily found. Likewise, online papers that
have been repeatedly viewed and downloaded, but never cited, may have more
public value than a cited paper with less internet traffic. Additional shortcomings are similar to traditional econometric approaches: what is the appropriate lag time for publications and what time window should be considered for
counting publications based on the depreciation rate of scientific knowledge
(Adams & Sveikauskas 1993)?

Scientists love to create models for complex problems, so it should be no surprise that a model was created that estimates the ultimate number of citations
for a particular article (Wang, Song & Barabási 2013). The model included the
following characteristics: citations accrue faster to papers that already have
many citations, a log-normal decay rate for future citations, and a general factor
that accounts for the novelty and importance of a paper. However, the model
required 5 to 10 years of citation history to make projections, and the difficulty
of properly calibrating the model limited its utility (Wang et al. 2014; Wang,
Mei & Hicks 2014). Conversely, an extensive study that looked at 22 million
papers published over the timespan of a century in the natural and social sciences found that citation histories are mixed and unpredictable (Ke et al. 2015).
Some extreme papers, labeled 'sleeping beauties,' accumulated few citations for

decades and then suddenly peaked—presumably because an important application for the research occurred at a much later date. Likewise, some of the most novel papers tend to languish for years in less prestigious journals but are eventually recognized by other fields for their original contributions and eventually become highly cited (Wang, Veugelers & Stephan 2016). Generally speaking, using short-term citations as a metric for assessing research is a bad idea.

A similar non-monetary approach for measuring research benefits is to count the number of patent citations in a particular field (Griliches 1979; Jaffe, Trajtenberg & Henderson 1993; Ahmadpoor & Jones 2017). This method has the benefit of better assessing the practical value of research activities and capturing the technological innovation component of research that is likely to have high social benefit. However, this method also shares some of the drawbacks of the publication approach as well as a few unique drawbacks of its own. The economist Zvi Griliches observed that a US productivity peak in the late 1960s was followed by a decline in patents granted in the early 1970s and that both events were preceded by a decline in the proportion of GDP devoted to industrial research spending in the mid-1960s. Whether productivity and patents followed a 5- to 10-year lag behind research spending was difficult to determine given that among other factors, the number of patents per research dollar also declined during that time period, an energy crisis occurred during that time period, and other countries suffered similar productivity losses without the drop in research funding (Griliches 1994).

Fluctuations in patent generation may also be due to the national patent office itself. For example, the 2011 Leahy-Smith America Invents Act, which took effect in 2013, changed the unique US first to invent patent system to a more standard first to file system. This makes comparisons before and after the new system more difficult. Likewise, as Griliches noted, stagnant or declining funding for a patent office could limit the throughput of the department or prevent it from keeping up with growing patent application submissions. This very phenomenon appears to have occurred in the US since the innovation boom of the Internet age (Wyatt 2011).

Ultimately, patents remain a limited and non-representative measure of research benefits. There is poor correlation between patents and public benefit because most benefits come from a small subset of all patents and only about half of all patents are ever used and fewer are ever renewed (Scotchmer 2004). Also, not all organizations patent their inventions at the same rate because the value of a patent is distinct from the value of the invention (Bessen 2008). Pharmaceutical patents can be extremely valuable; whereas, software-related innovations are more difficult to defensibly patent and are often obsolete before the patent is even awarded. A review of the 100 most important inventions each year from 1977 to 2004, as judged by the journal *Research and Development* ('R&D 100 Awards'), found that only one-tenth were actually patented (Fontana et al. 2013). Most companies relied on trade secrets or first-to-market advantages rather than patents. Patents allow a holder to litigate against infringement, but this legal right is often too expensive and time-consuming for all but the largest organizations to carry out. Alternatively, a large collection of related patents can create a 'patent thicket,' where its primary value is rent-seeking and slowing competitors, not social benefit. A CDC list of the most important public health achievements of the 20th century contained no patented innovations (Boldrin & Levine 2008), suggesting patents are indeed a very poor measure of research social benefit. Nonetheless, patents are still widely used as a measure of research value for lack of a convincing alternative.

While economists view the measurement of knowledge output to be problematic but possible, others believe the problem is intractable or at least not quantifiable in any honest way. Philosopher Paul Feyerabend argued that a careful study of the history of science shows the truth or usefulness of any particular scientific theory or line of research may not be appreciated for decades or even centuries (Feyerabend 2011). He gave one extreme example of the theory proposed by the Greek philosopher Parmenides of Elea (5th century BCE) that all matter has the same fundamental nature. The theory was abandoned for over 2,000 years before being revived by particle physicists in the 20th century. A more recent example is the theory of continental drift, first proposed in 1596 by Flemish cartographer Abraham Ortelius. The theory was revived in 1912 by meteorologist Alfred Wegener, who unsuccessfully championed the idea for two decades.[6] After the steady accumulation of supporting evidence, the idea was eventually incorporated into the theory of plate tectonics in the 1960s, which now serves as a cornerstone of modern geoscience. Perhaps the most relevant example is the theory that fossil fuels cause global warming, which was first proposed by Swedish scientist Svante Arrhenius in 1896. His work was inspired by British scientist John Tyndall's 1859 work on the radiative heat absorption properties of carbon dioxide and water vapor and their likely effects on the planet's surface temperature. Despite winning the 1903 Nobel Prize in Chemistry for foundational work in electrochemistry, Arrhenius' work detailing the correct mechanism and mathematical relationship between infrared absorption and atmospheric carbon dioxide concentrations was largely ignored for almost a century before anthropogenic climate change was realized to be an unprecedented threat to humanity.

Even though economists are generally trying to measure the short-term societal benefits of more tangible and immediate research, selecting a lag time is merely a choice of analytical convenience. There were decades between the development of quantum physics and technologies based on quantum theory: transistors, lasers, magnetic resonance imaging, and so on. The theory is over a century old and yet new technologies, such as quantum computers, are still in

---

[6] His failure was partly scientific, his observations had no good explanatory mechanism, and partly social, he was an outsider to the geology community and a German World War I veteran.

development. It would be hard to argue that these were impractical or unimportant benefits that could be left out of a realistic benefits assessment. It would seem even a field of research that has yet to yield useful results—such as string theory (Castelvecchi 2015)—should not be dismissed as long as it still has intellectual inspirational value; one never knows what is yet to transpire. Likewise, how does one measure the benefits of long-term research that may require decades to yield significant findings (Owens 2013b).

Selecting a lag time by a cutoff function that is designed to capture most of the citations, patents, or economic growth based on past research is based on the questionable assumption that only the intended outcome of applied research is of interest. However, the history of technology suggests secondary unintended discoveries, both good and bad, are important. For example, in the pharmaceutical industry, drugs are commonly repurposed when they are unexpectedly found to treat a disease other than their intended target. Thus, selecting a time period for the evaluation of research may capture some of the intended outcomes but miss the secondary serendipitous discoveries (Yaqub 2018).

Valuation by multiple metrics

The various metrics discussed so far appear to be poor measures of the social benefits of research. They are popular primarily because they make use of the available data, not because they necessarily measure the desired outcomes. Metrics are frequently pursued with the noble intention of improving accountability and transparency but do not often accomplish either because they tend to oversimplify complex processes and create perverse incentives to game the system when metrics are used to reward or punish individuals.[7]

For example, if patents become a preferred metric of research productivity, some researchers will knowingly generate patents that are of questionable licensing value to improve their likelihood of securing future funding. Likewise, the frequent practice of using the number of publications as a metric has led to academic complaints about 'salami-slicing' research and jokes about the 'least publishable unit.' Quantitative assessments of research output in the United Kingdom, Australia, and New Zealand may have created the unintended consequence of pushing researchers away from high-risk basic research and toward more conventional, short-term, applied projects to improve their rankings (Owens 2013a; McGilvray 2014) History suggests abandoning basic research in favor of seemingly more predictable short-term applied research is

---

[7] The metrics-focused system analysis approach of Secretary of Defense Robert McNamara is often blamed for the tragic poor decision-making surrounding the war in Vietnam as well as his later missteps as president of the World Bank. Proponents of metrics often point to successes in far simpler and more quantifiable human endeavors, such as baseball (Muller 2018).

probably counterproductive. For example, how could one predict that the germ theory of disease developed in the 19th century would be the impetus for the modern sanitation techniques responsible for much of the increase in average life expectancy in the 20th century? A review of almost 30 years of biomedical research grants found that basic and applied research were equally likely to be cited in patents (Li, Azoulay & Sampat 2017). Of course, the underlying observation is not new. Abraham Flexner first made the argument that basic research yields important social benefits in his 1939 essay, *The Usefulness of Useless Knowledge*. It appears the message requires frequent repetition.

Despite these critiques, there has been some hope that using a family of complementary metrics would yield an improved estimate over individual research measurements. For example, a combination of publication citations to capture basic research and patents to capture technology development might appear to be a complementary set of measurements. The STAR METRICS program was created to measure the impact of US federally funded research using a multi-dimensional approach. Some of the proposed indicators included (Federal Demonstration Partnership 2013):

- number of patents;
- number of start-up companies;
- economic value of start-up companies over time;
- future employment of student researchers;
- impacts on industry from research;
- the number of researchers employed;
- publications and citations; and
- long-term health and environmental impacts.

While the STAR METRICS approach avoided some of the limitations of individual metrics previously discussed, it was questionable how many of the proposed metrics could be measured in practice or how representative the final set of metrics would be. Given the difficulty of the task, it was not surprising when the full implementation of STAR METRICS program was abandoned in 2015.

A more successful program, the Innovation Union Scoreboard, has been evaluating the research efforts of European Union member states since 2007 (Hollanders & Es-Sadki 2014). It encompasses 25 metrics, including multiple indicators for educational outcomes, scientific publications, patents, public research investments, private research investments, employment, and other economic indicators. As with similar programs, the Innovation Union Scoreboard is by necessity restricted to indicators for which there are data. As such, unquantifiable benefits are missed.

Despite the difficulties of quantitatively valuing research, the era of big-data has inspired an entire alphabet soup of research assessment systems, none of which can be easily compared to each other. Detractors have argued that these broad quantitative measurement tools are just as non-representative and easily gamed as the many popular, but widely derided, college ranking schemes. It has yet to be seen if any of these multi-metric systems will improve research or how—outside their own definition—success will be determined. However, the rush to quantitative assessment is not universal. The Chinese Academy of Sciences moved away from an existing 24 indicator multi-metric research ranking system to a qualitative system based on peer review (Kun 2015). The motivation was a desire to place emphasis on the real social value of research rather than on easily measured surrogates.

Value-of-information analysis

Econometric methods would appear to be the obvious choice for performing a research cost-benefit analysis. However, as previously discussed, this is a difficult task even for research that has already been conducted. Estimating the value of future research is even more uncertain as it requires the questionable assumption that the future will be much like the past. This is a difficult assumption to defend because history shows the progress of technology to be inconsistent and unpredictable. Computer technology has exceeded most predictions made in the 20th century, yet utilities powered by nuclear fusion have stubbornly remained a technology of the future. Unfortunately, there is no consistent set of criteria that will predict whether a particular research project will succeed. The list of contributing factors is extensive, and there is even disagreement among studies regarding the magnitude and direction of influence of each factor (Balachandra & Friar 1997).

For future research decisions, an alternative to traditional econometric or knowledge output approaches is to use value-of-information (VOI) analysis. In VOI, the value of the research is measured by estimating its expected value to a particular decision and weighing it against the cost of obtaining that information (Morgan, Henrion & Small 1990; Fischhoff 2000).[8] For example, knowing the transmissibility of a particular pathogen has value for public health officials in their decision of how to prepare for future pandemics. This value can be measured in any agreeable units—money, lives saved, response time, and so on. The primary strength of this approach is that it deals directly with the value of research to the decision maker (Claxton & Sculpher 2006). By comparison, high quality research, as measured by knowledge output methods, has no clear correlation to societal benefit only an assumed link. Because VOI is a forward-looking predictive method of valuation rather than a backward-looking

---

[8]   VOI literature often uses the term 'expected value of perfect information,' which is simply the difference between the value of the decision made with complete information compared to existing information. Restated, this is the value of removing uncertainty from the decision process.

reflective method, it sidesteps the issue of making comparisons between past and future research.

Another strength is that VOI analysis is a more theoretically complete and consistent method of research valuation. Performing a cost-benefit analysis using a family of economic, knowledge, and social metrics can use collected data, but that data will generally be an incomplete measure of the total value of research and will often consist of proxies for the characteristics we would prefer to measure. Conversely, a VOI approach can place a direct value on factors that are difficult to monetize: aesthetic, intellectual, or even the cultural significance of a scientific discovery. Thus, VOI is complete in the sense that any recognized benefit can be included in the analysis.

However, the thoroughness of the VOI approach comes at the price of subjective estimates and value judgments. VOI is a productive decision tool only when one can reasonably estimate the value of obtaining the information. For that reason, VOI is often applied to business, engineering, and applied science decisions (Keisler et al. 2013). For example, VOI would be useful for estimating whether a particular medical test has value for decisions about patient treatment. However, it is harder to use VOI for estimating the value of highly uncertain basic research. VOI is subjective when it measures subjective things. It cannot create certainty out of uncertainty.

The thoroughness of the VOI approach also complicates analysis due to the broader array of potential social benefits that might be considered. While VOI is simple in concept, it can be quite complex in practice. For this reason, the VOI approach is often used in conjunction with an influence diagram—a visual representation of a decision process that represents variables as nodes and interactions between variables as arrows (Howard & Matheson 2005). The influence diagram serves as a visual aid to elucidate and organize the often complex interaction of factors that can affect the value of basic research. However, an influence diagram with more than a dozen or so nodes and arrows tends to become an unreadable labyrinth that provides little insight.

As an example, Figure 1 shows the relation among the various ways in which research can be valued as an influence diagram. Each form of valuation is represented as a node with arrows indicating if the method informs another method. For example, job creation is often concurrent with economic growth (but not always), so we would expect these two research valuation methods to be closely related. Likewise, both jobs and economic growth can be used in a multi-metric approach or in an expert opinion approach. Knowledge output, in the form of citations and patents, can also be used in a multi-metric approach and is similar to the VOI approach in that both are non-monetary and can more easily characterize the value of basic research with no immediate practical applications. Expert opinion, discussed in the next section, is the most comprehensive approach in that it can make use of all the other methods of valuation. However, in practice, expert opinion can range from superficial to comprehensive.

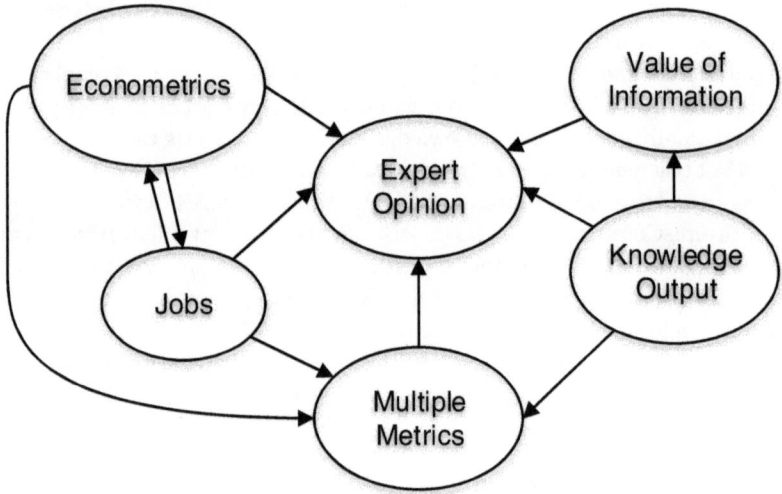

Figure 1: Methods of valuing research.

Although less formal than the VOI approach, a similar process can be used to reconcile the supply and demand for science research (Sarewitz & Pielke Jr 2007). This is done by collecting the information required by policymakers (the demand) through workshops, surveys, interviews, and committees. Using the same process, regular assessments are made regarding whether the research (the supply) is actually being used. Rather than placing a common value on the information, the intent is only to re-align research priorities to maximize social benefit. In theory, this is a great idea because useful science can happen by accident, but more useful science will happen when it is done with purpose. The priority re-alignment process is much less subjective than VOI in the sense that it does not attempt to quantitatively compare research programs. However, it is also time consuming in that it seeks input from all stakeholder groups and can be difficult to complete when contentious issues preclude a consensus on the demand for science. Furthermore, it is difficult to predict what research will actually yield the most social benefit; focusing only on immediate applied research would miss important basic research that eventually yields important technology.

In standard VOI literature, benefit is derived from additional knowledge, and it is assumed that the value of information can never be negative because a decision maker can always choose to ignore low-value information (Blackwell 1951). However, experiments suggest decision makers are often unable to ignore unhelpful information once it is known due to a 'curse of knowledge'

(Camerer, Loewenstein & Weber 1989).[9] Furthermore, decision makers are often unaware when information is unhelpful based on their surprising willingness to pay for unhelpful information (Loewenstein, Moore & Weber, 2003). This questions the basic assumption that the value of information is never negative because it can be ignored without cost.

We can extend this concept of negative value of information to include research that may yield knowledge that has potential public harm, such as dual-use research that has obvious use by military, terrorists, or criminals. Without the negative VOI concept, research cannot be any worse than wasted effort. With the idea of negative VOI, some research programs may yield information we might prefer not to know or find morally objectionable (Kass 2009). Likewise, some research might harm the public because it is erroneous. For example, a 1998 *Lancet* paper linked the MMR vaccine with autism. Although later discredited and retracted, it fueled suspicion regarding the safety of childhood vaccination; subsequent outbreaks of preventable diseases and multiple fatalities occurred in communities that disproportionately avoided vaccination (Gross 2009).

Qualitative assessment by expert opinion

A 1986 US Office of Technology Assessment report reviewed a variety of quantitative methods for determining the value of research and the prevalence of such methods in industry and government. The report found that the majority of managers preferred 'the judgment of mature, experienced managers' as the best method for assessing the value of research (OTA 1986). Formal quantitative models were perceived to be misleading due to their simplistic nature, which missed the complexity and uncertainty inherent in the decision-making process.

Given the issues with various quantitative methods as previously described, it is not surprising that expert opinion is still the gold standard in estimating the value of research. However, qualitative expert review is also problematic. Two fundamental difficulties with using expert opinion are conflicts of interest and unavoidable bias. Specialists are usually employed within their field of expertise, which leads to a weak, but pervasive, financial conflict of interest. Likewise, people tend to attach the most value to activities on which they have spent the most time. This phenomenon, referred to as effort justification

---

[9] An example of the curse of knowledge occurs in teaching. It is extremely difficult to imagine one's own state of mind before a concept was understood. This leads to teachers often overestimating the clarity of their instruction and the comprehension in their students (Weiman 2007).

(Festinger 1957) or the IKEA effect (Norton, Mochon & Ariely 2012)—because people tend to value an object more when they assemble it themselves—can lead experts to unintentionally overestimate the value of the research with which they have been most involved. Even the appearance of conflict between what is in the best interest for the general public versus the experts themselves decreases credibility and can make research assessment discussions look like special interest lobbying.

One way to partially compensate for potential expert bias is to actively seek competing views. Philosopher Philip Kitcher recommends an 'enlightened democracy' where well-informed individuals selected to broadly represent society set science research agendas (Kitcher 2001). This ideal is set as a middle ground between a 'vulgar democracy', where science suffers from the 'tyranny of the ignorant,' and the existing system, where a struggle for control over the research agenda is waged between scientists (internal elitism) and a privileged group of research funders (external elitism). Some influences on the science research agenda, such as focused lobbying by well-informed advocates, defy this idealized distinction between a scientific elite and an uninformed public. Nonetheless, the struggle to maintain a balanced and representative set of research policymakers is real.

One example of this struggle was the President's Science Advisory Committee created by US President Eisenhower to provide cautious science policy analysis during the American pro-science panic that occurred after the launch of Sputnik in October 1957. The Committee's criticism of President Kennedy's manned space program and President Johnson's and President Nixon's military programs led to its ultimate demise in 1973 (Wang 2008). The subsequent Office of Technology Assessment served in a similar role for the US Congress but faired only marginally better lasting from 1972 to 1995. It attempted to maintain neutrality by only explaining policy options without making explicit recommendations. However, its general critique of President Reagan's Strategic Defense Initiative—mockingly called Star Wars—created conservative antipathy that eventually led to its demise. Suggestions have been made on how to make such science advisory bodies more resilient (Tyler & Akerlof 2019), but these anecdotes suggest that balanced counsel on science policy can be difficult to maintain.

Compared to quantitative methods, assessment by expert opinion is time-consuming and expensive. The tradeoff is supposedly a better assessment. Unfortunately, the historical record is less than convincing For example, the National Science Foundation (NSF) uses peer review panels to assess research proposals based on the significance of goals, feasibility, the investigator's track record, and so on, but the process may not be capable of predicting even relative rankings of future research impact. For a study of 41 NSF projects funded a decade prior, panelists' predictions of future success were found to have no significant correlation with the actual number of publications and citations coming from each funded project (Scheiner & Bouchie 2013).

While expert panels are frequently used with the idea that group decisions are better than individual reviews, scientists are not immune to social dynamics that hinder good decision-making. Non-academics or other outsiders can be sidelined, dominant personalities or senior scientists may expect deference, or the panel may engage in groupthink. Larger studies of the NIH peer review process have found that there is no appreciable difference between high- and low-ranked grant proposals in their eventual number of publications per grant, number of citations adjusted for grant size, or time to publication (Mervis 2014). However, a study of 137,215 NIH grants awarded between 1980 and 2008 found that the highest-rated grant proposals yielded the most publications, citations, and patents such that a proposal with a review score one standard deviation above another generated 8 percent more publications on average (Li & Agha 2015). Critics have questioned the cause of this correlation considering journal publications are also based on peer-review; thus, any correlation may only indicate measurement of the same reputational system.

The journal peer-review system was the subject of another study that followed the publication history of 1,008 submissions to 3 top medical journals (Siler, Lee & Bero 2014). Of the 808 manuscripts that were eventually published, the lowest-rated submissions tended to receive the least eventual citations. However, the top 14 papers were all rejected at least once, which suggests the most innovative high-impact work is often unappreciated by the peer-review process.[10]

Perhaps the most damning critique of expert opinion comes from the many examples throughout history of substantial scientific research that went unappreciated by experts to an extent that is almost comical in hindsight. For example, biologist Lynn Margulis' paper proposing that mitochondria and chloroplasts in eukaryotic cells evolved from bacteria (Sagan 1967) was originally rejected by over a dozen journals. Over a decade later, DNA evidence confirmed the theory and Dr. Margulis was eventually elected to the National Academy of Sciences and given various awards, including the National Medal of Science. In another example, materials engineer Dan Shechtman needed two years to get his paper identifying the existence of quasicrystals published (Shechtman et al. 1984). He was met with ridicule from the scientific community and was even asked to leave a research group. This work eventually earned Dr. Shechtman the Nobel Prize in Chemistry in 2011.

The imprecision of peer review should come as no surprise to anyone who has published in the academic literature for some time. It is not uncommon to receive multiple reviewer comments that make contradictory assessments of a manuscript's quality or request mutually exclusive changes. Critiques of peer review have been common since its inception (Csiszar 2016), and there have

---

[10] While most rejections were desk rejections (that is rejections by the journal editors), this is, in practice, part of the peer-review process. This conservatism may be a sign of 'normal science' in action (cf. Kuhn 1962).

been many attempts to improve the process: abandoning anonymous reviews to improve accountability, publishing reviews to improve transparency, using double-blind reviewing to remove bias for or against the author's reputation or publication history, or even awarding grants by random lottery to proposals that meet established quality standards. Some of the calls for reform are in fundamental conflict with each other—some want to fund projects not pedigrees, while others want to fund people rather than projects—and each side has a plausible argument. While these changes may make the process fairer, it is unclear if they also improve the ability of experts to assess the long-term merit of research. Ultimately, we are left with the likelihood that expert opinion is the worst way to assess the benefits of research, except for all the other methods.

Implications for Assessing the Benefits of Research

A comparison of the most common ways policymakers assess the benefits of research provides some insight into science policy. Figure 2 shows the various approaches previously discussed ordered from the narrowest to the broadest conception of benefits. This also corresponds to ordering from the most objective to most subjective. That is, assessing the job creation potential of a research program is comparatively objective and data-driven, while expert opinion requires considerable use of subjective estimates and value judgments. The choice of approach depends on the purpose of the assessment. One can use these various methods to obtain answers that are either objective and incomplete or comprehensive and subjective but not objective and comprehensive.

For example, in the H5N1 virus case study in the previous chapter, the benefits of research using potential pandemic pathogens are highly influenced by social factors suggesting that a broader conception of benefits is more appropriate for assessment. Specifically, influenza research is most useful for regions that have functional public health systems. Given the uneven distribution of basic public health services in the world, any research benefits are far more limited in extent than in an ideal world. The problem is further exacerbated by

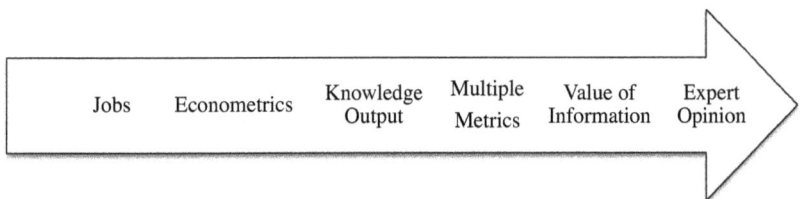

| Jobs | Econometrics | Knowledge Output | Multiple Metrics | Value of Information | Expert Opinion |

Figure 2: Ways of assessing the benefits of research ordered by increasing completeness of benefits that can be considered and also increasing uncertainty and subjectivity of estimates.

the frequent regression of public health services in regions experiencing war and failed governments. The sad reality is most people in the world have no access to an influenza vaccine of any kind. Since past influenza pandemics were not recognized in their early stages, the likelihood that tailored vaccines can be quickly distributed worldwide is small. These factors undermine the immediate practical health benefits of this research. While this nuanced view of science research benefits is useful, it increases the difficulty of quantification and the uncertainty of the assessment.

Upon reflection, we can see that some types of research are more amenable to particular forms of assessment. This suggests scientists involved in 'blue skies' basic research that has only job creation as an immediate quantifiable benefit should avoid getting locked into an econometric valuation debate. When basic science is treated as a mere economic engine, the weaknesses rather than the strengths of curiosity-driven research are emphasized, resulting in weak justifications. Rather, basic science should be honestly argued on intellectual, aesthetic, and even moral grounds if support from the general public is expected.

For example, in 1970, Ernst Stuhlinger, a scientist and NASA administrator, responded to a letter from Sister Mary Jucunda. Given the plight of starving children in Africa, she questioned the expenditure of billions of dollars for manned space flight (Usher 2013). Stuhlinger's response is an eloquent defense of the value of research in general but a rather weak defense of space exploration based on several proposed practical benefits—none of which are actually dependent on manned space flight: satellite data to improve agricultural output, encouraging science careers, increasing international cooperation, and serving as a more benign outlet for Cold War competition. However, Stuhlinger wisely closes the letter with a reference to an enclosed photograph of the Earth from the Moon and hints at its worldview changing implications. The 1968 picture, now referred to as 'Earthrise,' was later described by nature photographer Galen Rowell as 'the most influential environmental photograph ever taken' (Henry & Taylor 2009). Sometimes the greatest benefits cannot be quantified.

Implications for Research Allocation

There appears to be no method for assessing the benefits of research that is comprehensive, objective, and quantitative. This can make any research assessment process rather contentious if all the stakeholders are not already in agreement. Some science policy experts have suggested that the best science funding strategy is simply stable investment over time (Press 2013). The NSF estimated that over $1 billion was spent over a 40-year timespan on the search for gravitational waves. The result was a technical and intellectual achievement that yielded a Nobel Prize and a new sub-field of astronomy.

However, without the benefit of hindsight, it is hard to present a clear justification of what constitutes optimal research support. And without justification, proposed funding goals can appear arbitrary and claims of shortages or impending crises may be met with skepticism (Teitelbaum 2014; National Science Board 2016). While this advice rightly acknowledges that research budgets should not be based on the perceived viability of individual projects, it fails to resolve the question of selection. Should policymakers treat and fund all research requests equally?

Clearly, the general public does have science research priorities. A quick internet search of charities operating in the US yields dozens of charities that include cancer research as part of their mission but none for particle physics. The intellectual pleasures of discovering the Higgs boson in 2013 were real, but medical science, with its more immediate application to human health, attracts considerably more public attention. This exact allocation issue was recognized 50 years ago by philosopher Stephen Toulmin who wrote 'the choice between particle physics and cancer research becomes a decision whether to allocate more funds (a) to the patronage of the intellect or (b) to improving the nation's health. This is not a technical choice, but a political one' (Toulmin 1964). The purpose here is not to argue over whether medical research is more worthwhile than particle physics. Rather, it is to highlight how different methods of valuing research have ethical and pragmatic dimensions that effect science policy. A jobs-only valuation approach might prefer funding particle physics research for the many construction and engineering jobs it supports. Meanwhile, an econometric approach might prefer medical research based on historical growth rates in the pharmaceutical sector. Finally, a knowledge output approach might be ambivalent between the two options.

Of course, even with explicit consideration, the expression of public values[11] in science policy is not assured in the near term. For example, if a nation chose to scale back on 'curiosity' science, it is not clear that displaced scientists and engineers would necessarily start working on applied projects that would more directly minimize human suffering. Scientists and engineers are not fungible commodities, neither are they devoid of personal preferences regarding how they spend their time—research is not simply a zero-sum game. Likewise, public research funding is generally small compared to many other government expenditures, which may have considerably less societal benefit. In this century, the US federal budget for science research has been approximately one tenth of military spending. One can only imagine the benefits to humanity if those numbers were reversed.

William Press, president of the American Association for the Advancement of Science, stated '[a] skeptical and stressed Congress is entitled to wonder

[11] Public values can be defined as the ethical consensus of society on what constitutes the rights, freedoms, and duties of individuals, organizations, and society. This definition also acknowledges that public values are not necessarily fixed, monolithic, or entirely compatible with each other (for example, valuing both liberty and security) (Bozeman 2007).

whether scientists are the geese that lay golden eggs or just another group of pigs at the trough' (Press 2013). Questioning the social value of science was prevalent in the early 20th century (Bernal 1939), but this skeptical attitude about the US science community fell out of favor for several decades after Vannevar Bush rather successfully argued that science should be insulated from the political process (Bush 1945).[12] Nonetheless, research assessment and funding decisions have always been predicated on an expectation of societal benefit. The predominant belief has been that research funding directly translates into knowledge and innovation. The problem is determining exactly what those benefits are and what they are worth.

In summary, there is no universally acceptable method for assessing the benefits of research. This does not mean that assessing the benefits of science research is impossible or uninformative, only that formal quantitative benefits assessments should be used with extreme caution. Quantitative results that appear objective may be hiding a great deal of subjectivity. Failure to consider the limitations of each method risks letting the chosen method shape the goal of the assessment—the reverse of what constitutes good policymaking.

---

[12] See Guston (2000) for a more detailed discussion of the history of changing expectations of science.

Values in Risk Assessment

Having discussed how the benefits of research are assessed, we turn our attention to assessing risks. The field of risk analysis is often categorized into three main branches: risk assessment, risk management, and risk communication (Paté-Cornell & Cox 2014). Risk management, the process of deciding how to address risks, is widely understood to include subjective value judgments (Aven & Renn 2010; Aven & Zio 2014). Meanwhile, there is an idealized notion that good risk assessments are relatively free of values (Hansson & Aven 2014). However, despite our best efforts at quantitative rigor, the outcomes of risk assessments reflect the many value judgments implicit in the assumptions of the analysis (Ruckelshaus 1984; MacLean 1986; Shrader-Frechette 1986, 1991; Cranor 1997; Hansson 2013). Unfortunately, this common misconception about the nature of risk assessment means that pervasive but overlooked value judgments can transform seemingly objective assessments into stealth policy advocacy (Pielke Jr 2007; Calow 2014).

Policy analysis that is useful to stakeholders requires the clear identification of all significant assumptions and judgments (Morgan, Henrion & Small 1990; Fischhoff 2015; Donnelly et al. 2018). Delineating assumptions can give an analyst insight into how to minimize unnecessary assumptions and account for the remaining assumptions in a more transparent manner. However, this is tricky because there is no systematic list of value assumptions in risk analysis to consult. Even if there was such a list, it would be controversial—value judgments are difficult to recognize.

Rather than attempt the Sisyphean task of exhaustively detailing every possible value assumption, the intent here is only to discuss some of the most common and contentious value judgments to illustrate the inherent subjectivity of risk assessment. This is useful because subject experts attempting a risk assessment for potentially dangerous science may be in unfamiliar territory.

How to cite this book chapter:
Rozell, D. J. 2020. *Dangerous Science: Science Policy and Risk Analysis for Scientists and Engineers.* Pp. 29–56. London: Ubiquity Press. DOI: https://doi.org/10.5334/bci.c
License: CC-BY 4.0

Many scientists are trained to view subjectivity as a sign of incompetence or unprofessional behavior. This is an unnecessarily narrow view of science and a particularly unhelpful attitude in risk analysis. A general roadmap of value assumptions may make a more convincing argument that all risk assessments involve unavoidable and important value assumptions that, if ignored, decrease the credibility of a risk assessment and its utility in formulating public policy.

It should be noted that the following discussions of each topic are only brief introductions with relevant references for further exploration. Some topics, such as the treatment of uncertainty, can be rather technical. The point here is only to show that conflicting schools of thought exist for many of the considerations within a risk assessment.

Categorizing Assumptions

The term values is used here to mean a broad class of knowledge that exists between facts and accepted theory at one end of the spectrum and mere opinion at the other. Here, values are conclusions, based on the same set of available data, over which reasonable people might disagree. This definition describes values as having a basis in facts that are used to form reasons for preferring one thing over another (MacLean 2009). For example, preferring oysters rather than chicken for dinner because it tastes better is an opinion. Preferring oysters rather than chicken because it is healthier or more humane are values. Those preferences are values because they are based on some underlying facts— specifically, oysters have higher iron content than chicken and oysters have much simpler nervous systems—but the implications of those facts are open to interpretation and subject to varying degrees of public consensus.

One might claim any position that has a basis in fact can, in theory, be determined to be true or false by constructing a logical argument by Socratic dialogue or similar means. However, in practice individuals have taken positions (core beliefs, ideologies, schools of thought, etc.) from which they will not be easily dissuaded without overwhelming evidence that may not currently exist. This is not to say these value disputes will never be settled, just not yet. And here we run into the greatest challenge in risk assessment—the need to say something substantial about a decision that must be made in the present despite a state of approximate knowledge and uncertainty. The result is risk assessments rife with value judgments.

Value judgments in risk assessment can be categorized in multiple ways. One popular distinction is between two general classes of values: epistemic and non-epistemic (Rudner 1953; Rooney 1992; Hansson & Aven 2014). Epistemic value assumptions pertain to what can be known and how best to know it. Epistemological arguments, traditionally discussed in the philosophy of science, are nonetheless values in that they embody judgments of relative merit (Hempel 1960). Non-epistemic value assumptions (i.e., ethical or aesthetic values) typically deal with what ought to be or what is desirable or acceptable—keeping in mind that what is socially acceptable and what is ethical are not always the same thing (Taebi 2017). It is important to note the classification of value judgments is not static. Some epistemic value judgments, such as the choice of appropriate statistical techniques, may move along the spectrum of knowledge toward accepted theory as evidence accrues. Likewise, some ethical value judgments eventually become universal and are no longer a source of dispute. History is full of behaviors that were acceptable a century ago but are universally condemned today—as well as the reverse.

While the epistemic/ethical distinction has philosophical importance, it is less critical for risk analysts because, as discussed later, some assumptions can be justified by both epistemic and ethical reasons. Rather, organizing value assumptions by where they arise in the analysis process is more useful for instructive purposes. Momentarily ignoring that the process is iterative, the following is a discussion of the value assumptions that arise in each step of a typical risk assessment.

Selection of Topic

Commonly, an analyst will be employed to conduct an assessment of a specific risk. However, when not already predetermined, the first value judgment made in any risk assessment is the choice of topic. When tasked with evaluating risk within a large pool of potential hazards, a screening mechanism or set of criteria is needed that involves some epistemic and/or ethical value judgments to prioritize efforts. People employ heuristics (i.e., mental shortcuts) when assessing commonplace situations that can lead to biases—risk assessment is no different (Tversky & Kahneman 1974; Slovic, Fischhoff & Lichtenstein 1980).

Because risk assessments are generally performed for situations that are perceived to be dangerous rather than benign, an analyst's perception of risk gives rise to important judgments on topic choice. For example, technological hazards are perceived to be more controllable than natural hazards, but also more dangerous and more likely (Baum, Fleming & Davidson 1983; Brun 1992; Xie et al. 2011). If the public perceives anthropogenic risks to be more threatening than natural risks, the result should be a tendency to conduct more technological risk assessments while overlooking equally or more risky natural hazards.[1]

Likewise, a risk assessment is often influenced by how easily we can imagine a risk—a bias known as the availability heuristic (Tversky & Kahneman 1973). Funding for a risk assessment often appears when a threat is fresh in the minds

---

[1] This phenomenon may be partially responsible for inadequate natural disaster preparedness. A related question is whether many nations allocate too many resources to military spending rather than natural disaster mitigation.

of a funding organization, rather than when the risk is the greatest (assuming actual risk is even knowable). Thus, assessments tend to follow current events. For example, newsworthy meteorite impacts and near-misses tend to periodically renew interest in assessing the risk of major impacts from near-earth objects. The value judgments involved are both epistemic and ethical. Epistemic in the sense that there is an assumption that the hazard is now more real and knowable and ethical in the sense that the hazard is now viewed as more worthy of analysis than other hazards.

Before we move on to other concerns, it is important to note that the term bias shall be used here as traditionally used among decision science experts who have made careers enumerating and explaining the various ways in which humans frequently make terribly illogical decisions when confronted with unfamiliar low-probability, high-consequence risks.[2] That said, decision science tends to focus on the negative anti-rational behaviors. However, some decision-making behaviors that fall outside of traditional rationalism can also improve decisions—moral duty, desire for autonomy, acts of altruism, and so forth. Bias is not always bad.

Defining System Boundaries

For any assessment, the boundaries of the analysis must be selected. This includes time frame, spatial scale, and relevant populations. While the boundaries are primarily determined by the topic, some epistemic value judgments are involved. First, the analyst must believe meaningful boundaries can be defined for a system. The idea of inherent interconnectedness has long existed in the philosophical traditions of Buddhism and Taoism but was an uncommon idea in Western science until the 20th century when the emerging field of ecology led to sentiments such as, 'When we try to pick out anything by itself, we find it hitched to everything else in the Universe,' (Muir 1911) and 'Everything is connected to everything else' (Commoner 1971). Nonetheless, the reductionist approach to science has enjoyed a great deal of popularity and success.

Risk assessments are often performed for natural and social systems for which an underlying hierarchical structure is not yet understood. Thus, the analyst is forced to make bounding decisions based on scientific (i.e., epistemic) judgments. Ideally, expanding an assessment to add secondary or tertiary factors will make incrementally smaller changes to the results thereby showing the assessment to be convergent, but this is not always the case. Practical limitations of available data and funds tend to dictate that a risk assessment be narrowly defined, yet synergistic effects, non-standard uses, or

---

[2] For a particularly readable summary of decision biases, see *The Ostrich Paradox* (Meyer & Kunreuther 2017).

other sociotechnical surprises can play a significant role in the overall risk of a new technology (Jasanoff 2016). To complete the assessment, the analyst must believe it is possible to know what factors can be left out without appreciably affecting the results.

Ethical and epistemic considerations also pervade many choices regarding the relevant population in a risk assessment. Should an assessment be restricted to humans or should it include all sentient animals or even an entire ecosystem? Should the study involve only current populations or include future generations? These value assumptions can be influenced by many factors, such as type of academic training. For example, one study found that physicists and geologists were more likely to believe that performing a risk assessment of a long-term nuclear waste repository was a reasonable task, while anthropologists and philosophers were considerably more skeptical (Moser et al. 2012).

Method of Assessment

A variety of decisions about how risk assessments are conducted can have profound impacts on the results. In addition to the usual debates about acceptable scientific methodology, there is also considerable variation in the risk conceptual frameworks that have been adopted by different organizations, countries, and academic communities (Clahsen et al. 2019). The following highlights some of the more controversial choices that must be made.

Unit of assessment

An analyst must decide whether the risks will be expressed in monetary units or another unit relevant to the risk, such as injuries/year. Non-monetary units of risk are commonly used in pharmaceutical, healthcare, and epidemiological risk assessments. However, it becomes more difficult to compare these assessments with non-medical priorities and to make policy recommendations that consider financial resources. Units are also important in that a different unit can change the perception of the risk (Wilson & Crouch 2001). Depending on the situation, a relative risk can appear to be much larger than an absolute risk or vice versa. For example, a WHO report regarding the 2011 Fukushima nuclear accident estimated that the lifetime risk of thyroid cancer for nearby infant females increased from about 0.75 to 1.25 percent. Depending on the desired effect, media reported the findings as a 70 percent relative rate increase or a 0.5 percent absolute rate increase. Subsequent mass-screening of Fukushima children for thyroid abnormalities resulted in many unnecessary medical procedures and considerable public anxiety (Normile 2016). Conversely, in the H5N1 avian influenza research debate discussed in the first chapter, one critique of the conducted risk-benefit assessment was that the consultant used relative probabilities rather than absolute probabilities, which appeared to understate or at least obscure the risk.

These choices are intertwined with risk communication and can have serious implications. A risk can be presented (intentionally or not) in such a way as to minimize or inflate the severity of a particular hazard. The influence of context—also known as framing effects (Tversky & Kahneman 1981)—is substantial. A basic aspiration of risk assessment is to provide unbiased information, so it is generally considered unprofessional to use framing to make a risk assessment appear more favorable to an analyst's preferences. However, there is widespread support for framing risk outcomes that encourage environmental responsibility, prosocial behavior, or positive health practices (Edwards et al. 2001; Gallagher & Updegraff 2012). So it seems that framing is bad, unless it is used with good intentions. How sure are we that someday those good intentions will not be seen as socially or scientifically misguided? Clearly, value judgments in risk assessments are not only pervasive, but also complicated.

Selection of a unit also involves an epistemic value judgment regarding the measurability of a characteristic. Presumably, an analyst would not pick an unmeasurable unit. The choice of units is also accompanied by important, and often unwitting, ethical assumptions. For example, the units of lives saved, life years gained (LYs), quality adjusted life years (QALYs), and disability adjusted life years (DALYs) all preferentially benefit different populations (Robberstad 2005). Likewise, the disparate nature of various risks and benefits often requires the use of assumption-laden conversion factors and equivalencies to make comparisons. Ideally, an analyst should present an assessment using multiple units to give perspective and aid comparisons, but this requires more time and effort.

Value of life

The primary benefit of using a monetary unit is the ability to integrate the results into a larger economic analysis. However, the conversion to monetary units requires some important value assumptions. The most controversial is the need to monetize the value of life. Attempts to quantify the value of life often use willingness-to-pay measurements or expert opinion. A defense of these estimates is that the number does not represent the worth of an actual life but rather the rational amount society should be willing to spend to decrease the probability of a death when individual risk is already low. However, measurement techniques can confuse willingness-to-pay with ability-to-pay—a much less ethical measure because it undervalues the lives of the poor. Likewise, individual behavior is inconsistent; the amount an individual will pay to *avoid* a risk often differs from the amount the same individual must be paid to *take* a risk (Howard 1980). Furthermore, an individual's willingness-to-pay to save a life appears to vary depending upon whether the choice is made directly or indirectly through a market (Falk & Szech 2013).

Willingness-to-pay methods have two other fundamental difficulties (MacLean 2009). First, public well-being and willingness-to-pay are not always equivalent. Some individuals have preferences that are counter to the well-being of society or even their own well-being. Second, economic valuation is an incorrect way to measure many abstract and essential values, such as duties associated with religion or community. Individual behavior is often inconsistent when people are asked to put a price on deeply held values, and such requests are frequently met with moral outrage (Tetlock 2003).

Discount rate

Another contentious issue regarding the use of monetary units is the choice of discounting rate or how much future money is worth now. One can choose a high discount rate, such as seven percent—the average return on private investment; a low discount rate, such as one percent—the typical per capita consumption growth; or even a discount rate that declines over time to account for the uncertainty of future economic conditions (Arrow et al. 2013). The economic implications of selecting a discount rate are complicated enough, but the discount rate is also a proxy for complex intergenerational fairness issues—how we account for a future generation's values, technology, and wealth. Selecting higher discount rates minimizes future costs and tends to place more burden on future generations. The time range of the assessment determines the importance of the discount rate. For example, the choice of discount rate is often the most fundamental source of disagreement in long-term climate change economic risk assessments (Stern 2006; Nordhaus 2007; Stern & Taylor 2007).

The public commonly minimizes the value of future lives (Cropper, Aydede & Portney 1994). Even the basic idea of ethical consideration for future generations is not universally accepted (Visser't Hooft 1999). However, international laws increasingly acknowledge that physical distance is no longer an excuse for exclusion from risk considerations; eventually, separation in time may no longer be an acceptable reason for 'empathic remoteness' (Davidson 2009).

Other methodological considerations

One of the first methodological decisions is whether to perform a qualitative or quantitative risk assessment. Selecting a qualitative assessment may indicate deep uncertainty, lack of confidence in available data, or even mistrust of available quantitative methods. For example, the FDA issued new guidelines in 2013 that rejected purely quantitative risk-benefit assessments for new drug approvals because they often leave out important factors that are difficult to quantify. Likewise, a quantitative assessment may be selected for good reasons, such as strong past performance, or bad ones, such as to use numbers to imbue an assessment with an air of authority. The academic prestige associated with mathematical analysis contributes to 'a frequent confusion of mathematical rigor with scientific rigor' (Hall 1988). There is also a general bias against qualitative work in many scientific fields that might steer an analyst toward a quantitative assessment to avoid being accused of speculation.

The form of assessment will also depend on the exact definition and treatment of risk used. There are multiple common meanings of the term risk: an undesirable *event*, the *cause* of an undesirable event, the *probability* of an undesirable event, the *expectation value* of an undesirable event, or a *decision* made with known probability outcome (Möller 2012). These definitions of risk range from vague to precise, and their frequency of usage varies by academic discipline (Althaus 2005). The public tends to use risk qualitatively and comparatively; whereas, decision theorists and economists are more likely to use the more quantitative definitions (Boholm, Möller & Hansson 2016). When the public uses a more expansive conception of risk, professionals may often dismiss public sentiments as a biased overestimation of a small risk (Aven 2015). However, even experts tend to use the term inconsistently, which adds to the confusion.

In the relatively young field of risk analysis, the expectation value interpretation of risk has become widespread along with the use of probabilistic risk analysis. In practice, this simply means risks are compared by combining events with their probability of happening. For example, in a hypothetical game of chance where you win or lose money, there could be a 50 percent chance of losing \$1 and a 50 percent chance of winning \$2. The expected utility of the game would be

$$0.5 \times -\$1 + 0.5 \times \$2 = \$0.50$$

If you played the game repeatedly, you would, on average, expect to win 50 cents per game. Expected utility, part of rational choice theory, can be a handy method of comparing options in many economic decisions. Unfortunately, expected utility can be an unhelpful way to express events that are infrequent and serious—a common occurrence in risk assessments (Hansson 2012). For example, using the previous game, what if instead there was a 50 percent chance of losing \$10,000 and a 50 percent chance of winning \$11,000. In this case, the expected utility would be \$500. That is a much better expected utility than the original game, but fewer people would be willing to play the second game. The reason is explained by prospect theory (Kahneman & Tversky 1979), which argues that people have an aversion to large losses that causes deviations from rational choice theory. The larger the loss, the less likely people will attempt to maximize expected utility.

Additionally, risk assessments frequently include important ethical considerations missed by expected utility. For example, a 10 percent chance of one death has the same expectation value as a 0.01 percent chance of 1,000 deaths.

However, few people would say these two possibilities are morally equivalent. Despite these obvious issues with expected utility, it is still commonly used in risk assessments.

A related concern is with the criterion of economic efficiency—an assumption used when attempting to maximize expectation values. Maximizing the utility for the most likely outcome seems like a good idea, but it may come at the cost of potentially missing less efficient, but more robust options (Ben-Haim 2012). There is an ethical decision here. Is it better to favor efficiency to avoid waste or is it better to favor robustness—options that work over a wide range of situations—in order to minimize worst-case scenarios? These competing values are frequently encountered elsewhere in life. For example, should one invest in stocks with the highest rate of return or in less lucrative, but more stable investments? The public wants airplanes and bridges to be built economically to minimize costs and avoid wasting resources but with a large enough safety factor to protect the lives of those using them.

In the case of extreme uncertainty or ignorance, it is pointless to attempt to maximize expectation value. Rather, analysts should encourage 'robust satisficing' (Smithson & Ben-Haim 2015), qualitatively optimizing against surprise. This entails retaining as many acceptable options as possible (while also avoiding indecision) and favoring options that are the most reversible and flexible.

Treatment of Error

In any scientific statement or statistical test, there are two types of errors that can be made. A Type I error is finding an effect or phenomenon where it does not exist (incorrectly rejecting a null hypothesis). A Type II error is failing to find an effect that does exist (accepting a false null hypothesis). Traditionally, scientists have focused on Type I errors because the emphasis is on avoiding the addition of false theories to the corpus of science (Hansson 2012). However, an emphasis on rejecting false positives will likely miss real hazards for which there is not yet conclusive evidence. This is the reason why some products are banned many years after they are first introduced—it may require considerable data to build a case that will survive Type I error rejection. In risk assessment and public policy, concentrating on Type II errors, false negatives, may be preferred (Douglas 2000). That is, the epistemological values of traditional scientific inquiry may not be appropriate for risk assessments. This suggests the need to select different criteria of what constitutes acceptable evidence. However, it is not always as simple as reanalyzing data with different criteria for rejection. One approach is to use the precautionary principle, which can be viewed as a qualitative attempt at minimizing Type II errors within a body of science that was generated using Type I error minimization criteria. However, detractors of the precautionary principle believe an emphasis on Type II errors strays too far from defensible science (Sunstein 2005).

Model Selection

Risk assessments require selecting a risk model, and the selection process always involves value judgments. For example, simply fitting data to a standard dose-response curve can be accomplished by a variety of similar statistical techniques: maximum likelihood, non-linear least squares, piecewise linear interpolation, and so forth. Selecting the method is an epistemic value judgment, and its effect on the outcome of the analysis may or may not be trivial. For example, both log-normal and power-law distributions are highly right-skewed (heavy-tailed) probability distributions that yield more extreme large events than normally-distributed phenomena, but distinguishing between the two when fitting data can be difficult (Clauset, Shalizi & Newman 2009). However, the distinction can be important because log-normal distributions have a well-defined mean and standard deviation; whereas, power-law distributions sometimes do not and their approximated average is dependent on the largest estimated or observed event (Newman 2005). This could substantially affect the results of a risk assessment that uses mean estimates of hazard magnitude (Hergarten 2004). This behavior is particularly relevant to pandemic risk assessments because the size of various epidemics may follow power-law distributions without finite means, including cholera (Roy et al. 2014), measles (Rhodes & Anderson 1996), and early-stage influenza (de Picoli Junior et al. 2011; Meyer & Held 2014).

Other decisions are a mix of epistemic and ethical. For example, when selecting a low-dose exposure risk model, an analyst might choose a linear or non-linear model (Calabrese & Baldwin 2003) based on an epistemic value judgment. However, the selection may also be an ethical value judgment reflecting the analyst's belief regarding whether a model should strive to be as scientifically accurate as possible or whether it should err on the side of being conservatively protective (Nichols & Zeckhauser 1988; MacGillivray 2014).

As another example, epidemiologists may choose to use a continuous or discrete mathematical model to represent the behavior of an epidemic. A discrete model is useful in that it is easier to compare to epidemiological data (which is usually collected at discrete times) and is easier for non-mathematicians to use (Brauer, Feng & Castillo-Chavez 2010). However, discrete models sometimes exhibit very different dynamics than the continuous models they are intended to duplicate when used outside of a narrow set of parameters (Mollison & Din 1993; Glass, Xia & Grenfell 2003). Thus, researchers must make judgments regarding the relative value of correctness versus tractability when selecting models. Some of the same concerns arise when selecting between an analytical model and an approximated numerical model.

Theoretical versus empirical choices

The basis of a risk assessment model may be primarily theoretical or empirical, although this distinction is somewhat artificial. Because risk estimates involve projections into the future, they all have a theoretical component—even empirical trend extrapolations rely on the theory that future behavior will mimic past behavior. Nonetheless, there are special considerations for models that are more empirical or theoretical in nature. For example, defining and characterizing uncertainty in theoretical models requires more assumptions because most uncertainty characterizations rely on statistical techniques which require data. Likewise, empirical models must assume the data used to create the model are representative of the full range of possibilities (Lambert et al, 1994). In *The Black Swan*, Nassim Taleb illustrates this data incompleteness problem with a parable about a turkey that believes the farmer who feeds him daily is his best friend—until Thanksgiving eve when the farmer kills the turkey without warning. The last data point turned out to be the important one.

Opinions on empiricism constitute an important value assumption that rests on the perceived relationship between data and theory. A simplistic conception of science is that observations are used as a basis to form theories and subsequent observations then support or refute those theories. However, facts and observations are theory-laden (Feyerabend 1975; Mulkay 1979; Rocca & Andersen 2017). While theories are often inspired by observations, these observations are unconnected until interpreted within a theoretical framework. Both data and theory are intertwined. Detailed observations of celestial motion by ancient astronomers were hindered from providing more insight by a persistent theory of geocentrism. Misinterpreting a lot of data with the wrong model only improves the precision of the error. In the Thanksgiving turkey parable, all the data was leading toward the wrong conclusion because the turkey misunderstood the essential relationship between turkeys and farmers.

There is a growing emphasis on empiricism in the era of Big Data, but data mining is helpful only if the appropriate data are analyzed with the correct theoretical interpretation. Likewise, claims that we now live in an era of data-driven science are only partly correct. It appears that theory is often trying to catch up with the mountains of data science produces, but no one collects and analyzes data without at least a simple implicit underlying theory. The 2008 global financial crisis occurred despite (and perhaps because of) countless financial risk models with copious data that failed because they were missing key theoretical dependencies. The use of big data has also been turned to the 'science of science' to explore the predictability of scientific discovery only to find fundamental unpredictability in the absence of an underlying theory (Clauset, Larremore & Sinatra 2017).

Likewise, it is a value judgement to dismiss non-empirical methods for generating scientific knowledge or assessing risks. The *Gedankenexperiment* (thought experiment) has been used with great success in many fields ranging from philosophy to physics. Sadly, not all fields of science yet appreciate the importance of theory-building as a complement to, rather than a component of, fact-gathering (Drubin & Oster 2010).

Level of complexity

Selecting the level of model complexity in a risk assessment entails some value judgments. Analytical complexity is often equated with thoroughness and appropriate representation of reality. Meanwhile, proponents of simplicity will invoke Occam's razor—the principle that simpler explanations are preferred. A common theme in modeling is that the ideal model is as simple as possible while still aiding in informed decision-making (Vezér et al. 2018). But what is the appropriate level of detail?

Analysts face the competing goals of representativeness and usefulness. A broad and comprehensive assessment may offer a nuanced description of risk, but such completeness may not lend itself to the clear comparisons needed for a policy decision. Level of complexity is usually a tradeoff. While a simple assessment may be easier to explore and explain, it runs a higher risk of missing critical relationships. Meanwhile, a complex assessment has a better chance of capturing all the salient components of a system, but it is also harder to evaluate, understand, and compare to competing assessments (von Winterfeldt & Edwards 2007). There are methods, such as hierarchical holographic modeling (Haimes 1981; Haimes, Kaplan & Lambert 2002) and fault tree analysis (Vesely et al. 1981), which can help enumerate all of the potential interactions within a complex system, but no systematic guide for inductively (bottom-up) or deductively (top-down) investigating risk can guarantee completeness.

One might presume complex systems generally require more complex analysis. However, simple modeling may be desirable when there is a lack of theory or when complex models are known to lack predictive ability. For example, the time between large earthquakes appears to follow an exponential distribution. Because this process is memoryless (the time between events is independent of previous events), simple average time between events is just as useful as a more complex model (Cox 2012a). A similar issue can arise when successful methods are applied to new situations. Seeing the success of complex quantitative modeling in the engineering sciences, analysts sometimes overreach by applying the same techniques to poorly-understood complex systems (Pilkey & Pilkey-Jarvis 2007). An approach that works for buildings and bridges may not work when modeling ecological systems.

There is also good reason to be cautious about the lure of nuanced models and theories (Healy 2017). While the general consideration of nuance is laudable, there are several nuance traps that theorists frequently fall into: the urge to describe the world in such fine-grain empirical detail that it provides no generalizable insight; the tendency to expand and hedge a theory with particulars in such a way as to close off the theory to rebuttal or testing; or the desire to add nuance merely to demonstrate the sophistication of the theorist. All these forms of nuance would result in a more comprehensive, but also less useful, risk assessment. In general, generative or exploratory work should emphasize foundational issues and insight over nuance; whereas, explanatory and evaluative work is necessarily narrower and more detailed.

Although the goals of the assessment and the subject investigated should dictate the level of complexity, there is plenty of subjective flexibility for the analyst. As previously mentioned, mathematical complexity tends to imbue an analysis with an air of legitimate sophistication even when the rational basis for the numerical valuation is weak. Conversely, explanations that are simple, symmetrical, or otherwise clever are often considered more elegant and preferable to more cumbersome explanations (Hossenfelder 2018). It is important to recognize that these are essentially aesthetic value judgments.[3]

Data Collection

Value judgments exist throughout the data collection process. Many of these happen outside the control of the risk analyst due to widespread publication bias (Young, Ioannidis & Al-Ubaydli 2008). Available data are limited by many factors, including the common value judgment that null results are not valuable information (Franco, Malhotra & Simonovits 2014), the preferential over-reporting of false positive findings (Ioannidis 2005), the apparent bias against research from developing nations (Sumathipala, Siribaddana & Patel 2004), the bias toward publishing already distinguished authors (Merton 1968), or the difficulty of finding research not published in English—the lingua franca of international science. The bias against negative findings is particularly widespread because private sponsors of research will often discourage investigators from publishing negative results for financial reasons. This can result in a published paper showing, for example, the efficacy of a drug despite multiple unpublished studies showing no effect that the scientific community and public never see. An effort to address the issue has been made in the medical community by creating a database of clinical trials before they start, but the problem is pervasive and longstanding—the bias for positive effects was first discussed by Francis Bacon in the 1620 *Novum Organon*. There is also increasing awareness that the widespread journal publication bias toward original findings is a primary cause of an existing reproducibility problem. If researchers are discouraged from ever treading old ground, how can we be sure that established science is really established?

Counterintuitively, the bias toward original findings is also accompanied by a bias against novelty. That is, there is a preference for results that are new, but only incrementally so. This conservatism is partly propelled by a hypercompetitive research funding market that tends to reward researchers who can prove they have already successfully performed similar research (Alberts et al. 2014). Another factor is the idea famously summarized by physicist Max Planck that science advances one funeral at a time. Like all social endeavors, science has a

---

[3] There may be a natural human aesthetic predisposition for reductionism—a technique successful in both science and abstract art (Kandel 2016).

hierarchy, and there is a general tendency to repress work that contradicts the views of the most eminent scientists in a field—until they graciously accept the new theory or 'retire' (Azoulay, Fons-Rosen & Zivin 2015).

That said, many value decisions are made by the analyst.

Data screening

Analysts make a variety of value judgments regarding what data should be incorporated into a risk assessment. Data screening criteria can include relevance of search terms, reputational selection of sources, adherence to particular laboratory practices, findings with a certain statistical significance, or surrogate data similarity to the risk in question (MacGillivray 2014). What counts as relevant data can be a particularly contentious epistemic value judgment based on widely-held reductionist views of causality—dating back to philosopher David Hume—that tends to exclude some forms of otherwise compelling evidence (see Anjum & Rocca 2019). These subjective decisions are one reason why meta-analyses, studies that quantitatively combine and assess prior research on a subject, frequently come to contradictory conclusions.

While some criteria, such as statistical significance, are primarily epistemic value decisions, others are largely ethical. For example, it is common to treat potential harmful effects below the current detection limit as acceptable risks (Hansson 1999). While this may be pragmatic, it rests on the assumption a low-dose threshold is well-established and uncontroversial or simply what you do not know cannot hurt you. An even more general bias in data screening is the tendency by experts to ignore evidence that has not been transmitted through the scientific community. Paul Feyerabend frequently noted the hubris of scientists who ignored the accumulated practical knowledge of history and local experience. But what should a risk analyst do? It is troublesome to assess the quality or get general acceptance of knowledge that has not already passed through peer review. Sometimes the best one can do is to actually look for indigenous knowledge and note when it contradicts accepted research.

Similarly, some science policy experts (Sarewitz 2016; Martinson 2017) argue the scientific community is generating too much research of poor quality, which makes finding the valid research all the more difficult. Their solution is to encourage scientists to publish more thoughtfully and less often. While it is hard to argue with the thoughtful part, the implementation sounds troubling. Yes, a great deal of research is not high quality, but—as discussed in the previous chapter—only time and many eyes can tell if any particular research has value. Ironically, publishing is still the best way to disseminate information and improve the corpus of science.[4]

---

[4] I also disagree with the very idea of 'too much research.' We are not running out of scientific questions, so any increase in the number of scientists is generally positive.

Data for rare events

Some risk events are so rare that little to no data are available. Several limited approaches have been used to address rare event risks, including statistical methods, such as bootstrapping (Efron & Tibshirani 1993); expert opinion and related variations, such as the Delphi method; using analogous situations; and bounding with scenarios (Goodwin & Wright 2010). Historical records, such as old tsunami markers in Japan, are an excellent source for scenario bounding of natural risk events. Even many anthropogenic risks, such as bioengineering and geoengineering, have analogous natural events for comparison.

Sometimes the rarity of an event will lead a risk analyst to simply declare a risk non-existent—a rather bold epistemic assumption that should be used with caution. Although none are universally accepted, there are statistical methods for estimating the probability of events that have not occurred (Eypasch et al. 1995; Winkler, Smith & Fryback 2002; Quigley & Revie 2011). It is better to declare something improbable unless you are sure it is impossible.

One interesting subset of rare events is the low probability, high consequence risk—for example, a pandemic caused by avian influenza virus research. Low-probability events, as well as improbable, but unrefuted, catastrophic theories (Cirković 2012), tend to get considerable attention in the media. One possibility is that the public merely overreacts to the 'social amplification of risk' (Kasperson et al. 1988). Alternatively, the attention may be a judgment by the public that catastrophic risks should not be reduced to probabilistic expectation values (Coeckelbergh 2009). Furthermore, from an epistemic perspective, the probability of occurrence is actually the probability of the event conditioned on the theory being correct (Ord, Hillerbrand & Sandberg 2010). Thus, if the model is wrong, the rare event might not be as rare as experts estimate. Thus, public concern is, to some degree, a measure of their faith in experts.

Curiosity is a nearly universal human trait. All one needs is some training in skeptical analytical thinking to yield a scientist capable of doing meaningful science. Groundbreaking research can come from any imaginative, persistent, and/or lucky scientist. The idea of too much science is reminiscent of various claims that all the major discoveries of science have already been made (for example, science journalist John Horgan's 1996 book, *The End of Science*). Such arguments, dating back at least a century (Badash 1972), start with short-term trends based in facts and extrapolate to unsupportable conclusions. In short, who knows if what we claim to know is right, if some fundamental truths are undiscoverable, or what scientific revolutions we are capable of in the future? The only way to know is more science, more scientists, and more research.

Expert opinion

The last, but probably largest, source of value judgments in data collection is the use of expert opinion. If the assessment is to use expert opinion, value judgments occur not just in the opinions of the experts themselves, but also in the selection of the experts and the method by which expert opinions are combined. There is no consensus on the best methods for collecting and combining expert opinions (Hammitt & Zhang 2013; Morgan 2014, 2015; Bolger & Rowe 2015a, 2015b; Cooke 2015; Winkler 2015; Hanea et al. 2018). Furthermore, much has been said on the human imperfections of experts that affect their expertise (Laski 1931; Feyerabend 1978; Jasanoff 2016).

Accounting for Uncertainty

The treatment of uncertainty, which exists in all measurements and models, is, by definition, a fundamental issue for risk assessments. Whether deliberately or by accident, analysts make a variety of epistemic value choices on how to express and propagate uncertainty within a risk assessment.

Deterministic versus probabilistic

The easiest choice is often to temporarily ignore uncertainty and use a deterministic model to perform the risk assessment. A deterministic model can explore uncertainty by varying the model parameters and then building a range of scenarios. This at least gives the analyst a range of potential outcomes, albeit with no associated likelihoods. This approach limits the uses of a quantitative risk assessment but is useful when there is no defensible basis for quantifying uncertainty.

Objective versus subjective probabilities

Using the definition of risk as a probabilistic expectation value, the most popular option in risk analysis is to use a probabilistic model where parameters and output are represented as distributions. In this case, the analyst must decide if the probabilities will be based solely on data or if they will also include subjective probabilities (Hansson 2010b). The subjectivist approach to risk is increasingly preferred as it accommodates commonly used sources of uncertainty, such as expert opinion (Flage et al. 2014). This is often necessary when we consider trans-science—questions that are technically answerable by research but will not be answered because of practical limitations (Weinberg 1972). For example, if a particular failure risk for nuclear power plants was estimated to be 1 in a million per year, it would not be empirically verifiable because, with only about 500 nuclear power plants worldwide, that failure would, on average, occur once in 2000 years. These types of low-probability events are unlikely to be testable or known with much confidence.

Given the difference in confidence one might place on a probability computed from empirical data versus expert opinion, differentiating between empirical and subjective interpretations of uncertainty may be important—especially if these various sources of data are being combined within the same analysis (Doorn & Hansson 2011). In some models, second-order uncertainty (uncertainty about the uncertainty) is included, but the utility and interpretation of such efforts are still not universally accepted.

Hybrid probabilistic methods

A third approach is to distinguish between two types of probability: aleatory (normal variation within a population or process) and epistemic (lack of knowledge or incertitude). Some risk analysts account for aleatory uncertainty with standard probability distributions and epistemic uncertainty with other techniques. One option is to represent epistemic uncertainty with intervals that represent upper and lower bounds on possible values. This is done in probability bounds analysis or the more generalized method of imprecise probabilities (Walley 1991; Ferson & Ginzburg 1996; Weichselberger 2000; Ferson & Hajagos 2004).

Using such alternatives to probabilistic uncertainty is still uncommon for two reasons. First, the mathematical techniques are less familiar. Second, there is a tendency in decision theory to assume that all outcomes are reasonably well-known. This error of treating real-world uncertainty like the known probabilities found in casino games has been called the 'tuxedo fallacy' (Hansson 2009).

Non-probabilistic methods

A fourth approach is to use alternatives to (or extensions of) probability theory (Pedroni Nicola et al. 2017), such as evidence theory, also known as Dempster–Shafer theory (Dempster 1967; Shafer 1976, 1990), or possibility theory (Dubois & Prade 1988; Dubois 2006). Likewise, related issues of vagueness or ambiguity can be addressed by fuzzy set theory (Zadeh 1965; Unwin 1986). Newer risk analytic methods with novel treatments of uncertainty include info-gap analysis (Ben-Haim 2006) and confidence structures (Balch 2012). There are even semi-quantitative approaches that take into account the varying degrees of confidence we have in knowledge (Aven 2008). These methods can be used in risk assessments with alternative conceptions of risk. A traditional risk definition is hazard multiplied by probability. An alternative risk perspective defines risk as a consequence combined with uncertainty (Aven & Renn 2009; Aven 2010).

Uncertainty representation

Methods of representing uncertainty reflect an analyst's epistemological philosophy. This is in contrast to the moral uncertainty inherent in decision-making (Tannert, Elvers & Jandrig 2007). There is no current consensus on the best way to represent uncertainty, when to use one form over another, or what tools should be used to assess uncertainty (see Refsgaard et al. 2007). There is not even a consensus on the number of forms of uncertainty. Proponents of Bayesian statistics generally argue all uncertainty is a measure of belief irrespective of its source and probability is the best way to express it (O'Hagan & Oakley 2004). At the other extreme are classification schemes that organize uncertainty by location, nature, and source that can result in dozens of unique types of uncertainty (van Asselt & Rotmans 2002; Walker et al. 2003). In practice, uncertainty representation may be based on extraneous factors, such as familiarity, academic tradition, or ignorance of alternatives. Likewise, risk communication concerns might eclipse epistemic concerns (Tucker & Ferson 2008). For example, an analyst might prefer mathematical simplicity over a slightly more informative, but less understandable, method of treating uncertainty. It may also be a good choice to err on the side of simplicity in cases where increasing analytic complexity could obscure lack of knowledge—'a prescription that one's analytical formulation should grow in complexity and computational intensity as one knows less and less about the problem, will not pass the laugh test in real-world policy circles' (Casman, Morgan & Dowlatabadi 1999).

Model uncertainty

While parameter variability is often well-characterized in risk assessments, many forms of epistemic uncertainly are underestimated or ignored due to lack of appropriate methodology. Model uncertainty, also referred to as model form uncertainty or model structure uncertainty, is rarely acknowledged, and the few methods for quantitatively addressing this form of uncertainty are not universally accepted (Ferson 2014). Whether and how to account for model structure uncertainty is yet another epistemic value judgment. In some cases, model structure may be the most important source of uncertainty. For example, when five well-respected consulting firms generated groundwater pollution risk models based on the same field data, the result was conceptually unique models with no common predictive capability (Refsgaard et al. 2006).

Model structure uncertainty is more difficult to characterize than parameter uncertainty because it is based on a lack of knowledge, rather than natural variability, and is less amenable to probabilistic representation. Often, a model structure is selected from a range of possibilities and then the modeler proceeds to account for parameter uncertainty while treating the model structure as a given (Draper 1995). The following are some existing strategies for addressing model structure uncertainty.

Lumped uncertainty

Where copious data exists, the modeler can use a split data set to first select the parameters during the model calibration phase and then evaluate the entire model during the validation phase. Deviations between the model output and the second data set can be partially attributed to model structure uncertainty. The structural uncertainty can then be accounted for by either increasing parameter uncertainty until it also accounts for structural uncertainty (this occurs in inverse modeling methods) or by adding an explicit structural uncertainty term—irreverently known as the fudge factor. The lumped uncertainty approach assumes the available data are reliable and representative and the underlying modeled processes are stationary.

Sensitivity analysis with multiple models

One of the mathematically simplest methods of addressing model structure uncertainty is to create multiple models that address the range of possible model structures. Each model is evaluated separately, and the output of all the models is summarized as a set of results. This approach was used in early climate modeling projections. It has the benefit of being simple to understand, but it can become arduous if there are many models. Likewise, summarizing the many results and guessing the likelihood of particular model outcomes is less straightforward. Most importantly, this approach works on the substantial epistemic assumption that the full range of possible models has been described.

Monte Carlo model averaging

A Monte Carlo probabilistic model can account for multiple model forms by sampling each possible model and combining their results into a single probability distribution (Morgan, Henrion & Small 1990). The sampling can take advantage of resampling techniques, such as bootstrapping, and can even be weighted if some models are more probable. This method requires all possible model structures be identified by the analyst. Another concern is averaged distributions will likely underestimate tail risks—the low-probability, worst-case events that occur at the tails of a distribution. More fundamentally, some analysts object to the idea of combining theoretically incompatible models.

Bayesian model averaging

Bayesian model averaging is similar to Monte Carlo averaging in that it combines all the identified potential model structures into an aggregated output (Hoeting et al. 1999). In this case, the model uncertainty, both parameter and structural, are evaluated together. Bayesian averaging has the same limitations as the Monte Carlo approach: completeness concerns, averaging incompatible theories, and underestimating tail risks.

Bounding analysis

Using bounding analysis, the results for all the potential models are compared and an envelope is drawn around the entire set of results. The final product is a single bounded region likely to contain the correct model output. A benefit of this method is all possible model structures need not be identified only those that would yield the most extreme outcomes. While there is no guarantee the analyst will be able to identify the extreme model structures, it is a simpler goal than identifying all possible models. Bounding analysis also avoids the issues of underestimating tail risks and averaging incompatible theories—it propagates rather than erases uncertainty (Ferson 2014). Weaknesses include the inability to weight the credibility of individual model structures and the inability to distinguish likelihoods within the bounded region.

Unknown unknowns

Analysts frequently encounter situations where uncertainty is deep but recognized. A variety of methods are available for dealing with extreme uncertainty in risk assessments (Cox 2012b). However, sometimes we do not even recognize our own ignorance. The phrase 'unknown unknowns' is relatively new, but the underlying concept is ancient—in Plato's *Meno*, Socrates points out one cannot inquire about a topic with which one is wholly ignorant. By definition, we are ignorant of unknown unknowns, so we generally exclude them from risk assessments. Nonetheless, approaches for reducing ignorance have been proposed, such as imaginative thinking and increased public discourse (to be discussed in more detail in the last chapter) (Attenberg, Ipeirotis

& Provost 2015). While the uncertainty associated with ignorance is unquanti-fiable, acknowledging the limitations of our knowledge is a display of Socratic wisdom that improves risk communication (Elahi 2011).

While there is uncertainty inherent in all knowledge, formal risk assessments tend to make specific, often quantitative, claims regarding the level of certainty of knowledge. Thus, special attention is warranted regarding the caveats placed in an assessment that reflect the analyst's value judgments regarding what is known and knowable. An analyst may believe that an assessment has captured all the salient points worthy of consideration to a degree of accuracy and pre-cision that conclusively answers the question. This analyst is likely to present findings with few caveats. A more skeptical and humble analyst will add quali-fiers to assessments so readers do not over-interpret the results.

Comparing Risks

One reason to create a risk estimate is to compare it to other hazards or risk management options. The comparison process is full of value judgments. For example, one of the most common comparisons of a hazard is to natu-rally occurring levels of a potential harm, such as background radiation levels (Hansson 2003). Using natural exposure levels as a standard for comparison when there is scant reason to assume this constitutes an acceptable level of harm is a value judgment. However, the widespread use of sunscreen suggests the general public does not always find natural risks acceptable. Experiencing a level of harm by default does not imply technologies that subject us to similar levels of risk are acceptable (Fischhoff et al. 1978; Slovic 2000).

Along the same lines, public concerns regarding various technologies, such as synthetic food additives or genetically modified foods, are sometimes based primarily on the unnaturalness of the technology (Viscusi & Hakes 1998; Hans-son 2012). However, this assessment is based on the belief naturally occurring substances are safer. This is a reasonable assumption in the sense humans have co-existed with naturally occurring materials for a long time. This provides extensive experience helpful for forming judgments of safety. However, it is a naïve generalization to assume natural equals safe. For example, arsenic is naturally occurring in the groundwater of some areas, and it poses a larger public health threat than many anthropogenic water contaminants. The basic assumption that natural substances are benign is even ensconced in US regula-tions; the FDA has far fewer requirements for botanical medicines (sold to the public as nutritional supplements) than for synthetic drugs. Even equivalent harms caused by natural sources are perceived to be less scary than human-caused harms (Rudski et al. 2011). The source of this distinction appears to be the 'risk as feeling' (Loewenstein et al., 2001) model and 'affect heuristic' (Slovic et al. 2007) which suggest perceptions of risk are dependent upon the emotions associated with the risk.

Incommensurability

One of the most basic assumptions in risk assessments is the belief risks can be compared—even using the precautionary principle is an implicit comparison between the potential risk and the status quo. However, is it always possible to compare any risk? Are some risks incommensurable? Certainly, risks that are different in nature (such as health risks and risk of habitat loss) are difficult to compare (Espinoza 2009). Any such comparison requires the use of a common unit of measure, such as economic value, or equally controversial subjective rankings. However, even risks that appear to be of the same kind (such as all risks that could shorten a human life) can still be difficult to compare due to important ethical distinctions. Public rejection of quantitative risk assessments in the past may not be due to risk communication failures but rather to the failure of these formal assessments to account for ethical distinctions important to the public (Hansson 2005; Kuzma & Besley 2008). Some distinctions often ignored in quantitative risk assessments that do not share ethical equivalency include (Slovic 1987; Gillette & Krier 1990; Cranor 2009; Espinoza 2009)

- natural versus anthropogenic risks;
- detectable versus undetectable (without special instrumentation) risks;
- controllable versus uncontrollable risks;
- voluntary versus imposed risks;
- risks with benefits versus uncompensated risks;
- known risks versus vague risks ('ambiguity aversion' (Fox & Tversky 1995));
- risks central to people's everyday lives versus uncommon risks;
- future versus immediate risks; and
- equitable versus asymmetric distribution of risks (in both space and time). Similar justice issues arise when the exposed, the beneficiaries, and the decision-makers are different groups (Hermansson & Hansson 2007; Hansson 2018).

In each pair above, the first risk type is generally preferred to the second risk type. In practice, risks often fit multiple categories and can be ordered accordingly. For example, common, self-controlled, voluntary risks, such as driving, generate the least public apprehension; whereas, uncommon, imposed risks without benefits, such as terrorism, inspire the most dread. Thus, it is no surprise US spending on highway traffic safety is a fraction of the spending on counter-terrorism despite the fact automobile accidents killed about 100 times more Americans than terrorism in the first decade of this century (which includes the massive 9/11 terrorism attack). Ignoring these risk distinctions can lead to simplified assessments that are completely disregarded by the public. It is for this reason risk governance guidelines stress the importance of social context (Renn & Graham 2005). While the importance of these distinctions has been understood for some time in theory (Jasanoff 1993), there remains limited evidence of this occurring in practice (Pohjola et al. 2012). Because these ethical distinctions are often publically expressed as moral emotions (duty, autonomy, fairness, etc.), ignoring the emotional content of risk assessment decreases both their quality and the likelihood they will be followed (Roeser & Pesch 2016).

Risk ranking

Some risk assessments may also employ a risk ranking method in the final comparison. The ranking may be quantitative or qualitative, single or multi-attribute, and can take many forms including, letter grades, number grades, cumulative probability distributions, exceedance probabilities, color categories, and word categories. The ranking method selection is an epistemic value judgment with important risk communication implications (Cox, Babayev & Huber 2005; Cox 2008; MacKenzie 2014). Most risk rankings are simply based on probabilities and consequences, but rankings incorporating ethical dimensions, such as the source of risk (Gardoni & Murphy 2014), have been proposed. This generates a more nuanced, but also more subjective, form of ranking. For example, which is the most concerning: 1,000 deaths caused by heart disease, 100 smoking-related deaths, or 10 homicides? The question is almost meaningless when stripped of its context. Unfortunately, this is precisely what happens in a risk ranking exercise without accompanying qualitative descriptions.

A Value Assumption Roadmap

Given the myriad value assumptions discussed in this chapter, an aid is useful. The following list (Table 1) is organized chronologically in the risk assessment process so it can be used as a checklist for risk analysts. The list is not exhaustive in its coverage of potential value judgments, but it highlights common and contentious assumptions that, left unexamined and unaddressed, decrease the utility of a risk assessment.

The roadmap of value assumptions ignores some of the more uncontroversial values inherent in risk assessment as well as broader value discussions within science. For example, the general debate over what constitutes quality science (Kuhn 1962; Feyerabend 1970; Lakatos 1970) is itself a value-based argument: 'The ethos of science is that affectively toned complex of values and norms which is held to be binding on the man of science' (Merton 1942). However, it is not always clear what epistemic, ethical, or aesthetic values are considered to be uncontroversial or for how long they will remain uncontested (Hansson & Aven 2014). Maximizing happiness is a common goal in contemporary economic

Table 1: A summary process map of value judgments in risk assessments.

Step	Fundamental Value Questions
Selecting a topic	• How are hazards screened? • What heuristics are influencing choice?
Defining the boundaries	• What is an appropriate time, space, and population? • Holistic or component analysis?
Choosing the assessment form	• What unit will be used? • What is a life worth? • What are deeply held values worth? • What discount rate should be used? • Qualitative or quantitative? • Which definition of risk? • Maximizing efficiency or resiliency? • Focus on preventing false positives or false negatives?
Model selection	• Accuracy or precaution? • Theoretical or empirical? • Simple or complex model?
Data selection	• How is data screened? • How are rare events treated? • How is expert opinion used?
Accounting for uncertainty	• Deterministic or probabilistic? • Objective or subjective probabilities? • How is incertitude addressed?
Comparing risks	• Can the risks be compared? • Qualitative ethical distinctions? • Risk ranking?

analyses. It might even seem reasonable to think of it as an uncontroversial value. Yet, different eras and cultures have valued duty over self-interest.[5]

Example: Risk assessment of farmed salmon

To see the utility of the value judgments map, it helps to apply it to an actual risk assessment debate. In the following example, the original study found high concentrations of carcinogenic organochlorine contaminants

---

[5] For an overview of various ethical frameworks as they apply to risk assessment, see Rozell (2018).

in farm-raised salmon and concluded the risk of consumption outweighed the benefits (Hites et al. 2004). The analysis prompted a series of strong response letters. One letter pointed out even using a conservatively protective US Environmental Protection Agency (EPA) linear cancer risk model, the expected number of cancer cases from consuming farmed salmon was a fraction of the number of cardiovascular mortalities that would be avoided by eating salmon (Rembold 2004). Furthermore, the critique noted the quality of the data was not the same; the cardiovascular benefits data was based on randomized clinical trials, while the cancer risk data was based on less reliable observational studies and nonhuman dose-response models. The response from the original authors raised the possibility of other non-cancer harms (i.e., neurological) from contaminated fish consumption, as well as a reminder that the beneficial omega-3 fatty acids found in salmon were available from other non-contaminated dietary sources.

A second letter also compared the cardiovascular benefits of salmon consumption to cancer risks and additionally included a value-of-information analysis to argue any uncertainty in the risk and benefit data did not affect the assessment salmon consumption was beneficial (Tuomisto et al. 2004). For this reason, the letter went so far as to imply that the original study was non-scientific. Again, the response pointed out fish were not the sole source of dietary cardiovascular benefits.

A third letter questioned the EPAs linear dose-response model, citing additional data that suggested there were no carcinogenic effects at low exposures (Lund et al. 2004). The response by the original authors argued the additional study used a sample size too small to detect the estimated cancer risk. Furthermore, they pointed out the potential neurological and cancer risk is larger for young people due to bioaccumulation, while the cardiovascular effects primarily benefit older individuals, which suggested the need to distinguish populations.

Looking at the various critiques and responses, value assumptions were made corresponding to each step of the risk assessment process. Boundary value assumptions were made regarding what can be counted as a risk (cancer and neurological impairment) or a benefit (cardiovascular health) and whether sensitive populations (children) should be considered separately. Likewise, there were debates regarding the appropriate risk model (the EPA linear model or a no-effects threshold model); what data were worthy of inclusion in the assessment (observational studies, animal studies, and small sample size studies); and how important it was to account for uncertainty. Finally, there was an important value judgment regarding risk comparison: will people who potentially stop eating salmon substitute other foods rich in omega 3 fatty acids (for example, should salmon be compared to walnuts)? While analyzing the value assumptions in the various risk assessments does not resolve the underlying scientific questions, it does help clarify the arguments and provide insight into what fundamental issues should be explored further.

Example: The avian influenza research debate

Returning to the avian influenza research discussed in the first chapter, we can look at some of the risk assessments associated with the debate and see where the contentious value judgements arose. While the NIH requested both qualitative and quantitative risk-benefit assessments, the general belief among scientists was a quantitative assessment would be more credible. This was a dubious assumption given the wide range of results from previous attempts. For example, a simple probabilistic risk assessment from the biosecurity community estimated research with potential pandemic pathogens would create an 80 percent likelihood of a release every 13 years (Klotz & Sylvester 2012). An updated assessment estimated the risk of a research-induced pandemic over 10 years to be between 5 and 27 percent (Klotz & Sylvester 2014). Meanwhile, a risk assessment from the public health community estimated for every year a laboratory performed this type of research, there was 0.01 to 0.1 percent risk of a pandemic, which would result in 2 million to 1.4 billion fatalities (Lipsitch & Inglesby 2014). When these results were presented at an NRC symposium, Ron Fouchier, the head of the original controversial study responded, 'I prefer no numbers rather than ridiculous numbers that make no sense.'

Dr. Fouchier's subsequent risk assessment estimated a lab-induced pandemic would occur approximately every 33 billion years (Fouchier 2015b). Since this is more than twice the known age of the universe, his calculated risk was essentially zero. Dr. Fouchier noted there has been no confirmed laboratory-acquired flu infections nor any releases in decades and this supported his conclusion the risk is now non-existent. Critics of his assessment questioned his methodology and selection of evidence (Klotz 2015; Lipsitch & Inglesby 2015). Dr. Fouchier responded with the same complaints (Fouchier 2015a).

A similar question over appropriate evidence arose when proponents of gain-of-function research posited that the 1977 flu pandemic was caused by a vaccine trial or vaccine development accident rather than a research laboratory release (Rozo & Gronvall 2015). They concluded 'it remains likely that to this date, there has been no real-world example of a laboratory accident that has led to a global epidemic.' Critics have argued this nuanced position is largely irrelevant to the gain-of-function debate (Furmanski 2015). Because vaccine development is still a primary goal of gain-of-function research, a vaccine mishap is no less worrisome than a research lab release—an epidemic could occur regardless of which lab made the fatal error.

An even more fundamental values debate over what constitutes evidence arises over the essential purpose of the research. Genetic analysis of influenza viruses cannot yet predict the functional behavior of a virus (Russell et al. 2014; M et al. 2016). Likewise, a specific sequence of mutations in the lab are not guaranteed to be the most likely to occur in nature. This limits the immediate value of the research for practical applications, such as vaccine design. Critics of the H5N1 research argued the lack of immediate application was a reason not to perform the work. Proponents had a different view of the same facts. Not understanding the connection between genes and virus behavior was a gap in knowledge that must be closed and gain-of-function research was the best way to obtain this valuable information. As with many things in life, perspective is everything.

The avian influenza debate also included value-laden risk comparisons. For example, Dr. Fouchier argued most of the biosafety and biosecurity concerns raised about his work also applied to natural pathogen research, which was not restricted—thus, gain-of-function research should not be restricted either (Fouchier 2015a). Temporarily ignoring the possibility that the consequences of an engineered virus release might be much greater, the underlying assumption of this argument is natural pathogen research is relatively safe. But, as previously discussed, comparisons to nature are appealing but do not resolve the essential risk question. Instead, this argument could be flipped and interpreted as suggesting that research on naturally occurring pathogens may also require further restriction.

Identifying these value judgments is useful in the difficult task of assessing the low probability-high consequence risk of a pandemic. Given the infrequency of major influenza pandemics, Harvey Fineberg, former president of the US National Academy of Medicine, has noted, 'scientists can make relatively few direct observations in each lifetime and have a long time to think about each observation. That is a circumstance that is ripe for over-interpretation' (Fineberg 2009).

Is Risk Assessment Useful?

Risk analysis has become more challenging over time. For most of human history, the assessment and management of risk was simply trial and error. However, as the 20th century unfolded, the power and scale of technology grew, and it became evident that society needed to assess risks proactively. Science has also allowed us to recognize and measure more subtle hazards. Coupled with a general decreasing tolerance for risk in modern industrialized society, the challenges to the field of risk analysis are considerable.

Where risks and benefits are clear and certain, a formal risk assessment is generally unnecessary; the process is intended for contentious issues where the correct policy decision is less obvious. However, it is in these very situations where the outcome of a risk assessment is strongly influenced by the many inherent value judgments (often unknowingly) made by the analyst. So are risk assessments useful? The short answer is yes. Even though this review of subjective values in risk assessment could be interpreted as a critique of the process, it is important to dispel any notion that 'subjective' is a derogatory term or it is necessarily arbitrary or based on a whim. Subjective values can be deeply-held convictions with rational supporting arguments. Past neglect of value judgments, especially ethical considerations, has ignored the importance of emotional content in decision-making or treated it as merely coincidental (Roeser 2011). Appreciation of these value-laden assumptions as inherent to risk assessment improves both the process and the final assessment.

Risk analysts who understand they are not merely collecting and tallying the facts surrounding a policy question will greatly improve their work. Yet it is no simple feat to maintain a skepticism and awareness of one's own assumptions. If an analyst believes risk assessments produce objective answers, any assessment produced will understate its subjectivity and incertitude. Narrow risk assessments of well-understood phenomena with ample data might be uncontested, but formal risk assessments rarely resolve public debates regarding controversial science and emerging technologies. The various forms of value assumptions made in risk assessments are a primary reason for the common inability of scientific studies to resolve disputes. Astute stakeholders and policymakers intuitively understand the limitations of risk analysis, and any assessment that is overly conclusive will be dismissed as deeply biased or naïve. Considering and clearly discussing the value-laden assumptions in a risk assessment improves trust in the provided information by allaying concerns of hidden agendas.

We still do not fully understand how risk assessment can be used to build consensus and reach decisions (Aven & Zio 2014). However, honest attempts to account for value judgments can aid rather than hinder public trust in a formal risk assessment where the ultimate goal is to provide information that is both useful and credible.

Technological Risk Attitudes in Science Policy

Science and technology policy debates frequently call for risk-benefit assessments based on sound science. But what constitutes credible science is a contentious issue in its own right (Yamamoto 2012), There is some consensus over what good science should look like in theory but much less regarding what it looks like in practice or in individual cases (Small, Güvenç & DeKay 2014). Furthermore, the science underlying many controversies is sufficiently complex and well-studied such that competing sides can take their pick of defensible science with which to argue their position (Sarewitz 2004). The result is formal risk-benefit assessments usually fail to resolve debates over controversial technology. The detailed arguments laid out in the prior two chapters hopefully provide a convincing case that the legitimate subjectivity found in risk-benefit assessments is unavoidable.

Science and technology studies pioneer Sheila Jasanoff criticizes the modern attitude that difficult technological decisions are always resolvable with further research (Jasanoff 2007). She argues the most difficult parts of the decision process are often ethical and political—not scientific. Thus, delaying decisions for more fact-finding is often a form of wishful thinking. Calls to perform risk-benefit assessments for potentially dangerous technologies often fall into this category.

In science and technology policy, we frequently encounter debates where well-intentioned and reasonable individuals still arrive at very different conclusions. This leads to the obvious question underlying the subjective dimensions of science and technology policy: given the same data, *why* do reasonable individuals disagree?

In any decision process, it is believed individuals tend first to evaluate novel information using mental shortcuts (heuristics) that are replaced by more reasoned thinking as familiarity with the subject increases (Chaiken 1980; Kahneman 2011). The process of evaluating science research and new technologies

How to cite this book chapter:
Rozell, D. J. 2020. *Dangerous Science: Science Policy and Risk Analysis for Scientists and Engineers.* Pp. 57–76. London: Ubiquity Press. DOI: https://doi.org/10.5334/bci.d.
License: CC-BY 4.0

is no different (Scheufele 2006). The heuristic-to-systematic thinking model also applies to interpreting risk assessments (Kahlor et al. 2003). New information can influence an individual's attitude, but any pre-existing attitude also influences how new information will be interpreted (Eagly & Chaiken 1993). Individuals with positive attitudes about technology will tend to expect more benefits from new technologies, while those with negative attitudes will expect fewer benefits (Kim et al. 2014).

Many factors can influence attitudes about technology. The purpose here is to discuss the nature of technological risk attitudes that may account for why well-informed and reasonable people disagree about controversial research and emerging technologies. An outcome of this discussion is insight into the management of controversial research (to be discussed in the following chapter).

Technological Optimism and Skepticism

Although attitudes regarding technological risk exist along a broad continuum, for our purpose it is helpful to define two general categories: technological optimism and technological skepticism. This simplification is useful because we are primarily concerned with policy formulation and the binary outcome of whether an individual endorses or opposes a particular line of research or technology—even the ambivalent and indifferent, which may constitute the majority of the general public (Seidl et al. 2013), will eventually (and perhaps unwittingly) fall into one category or the other. The dual categorization has also been used by other academic and popular writers (e.g., Costanza 1999, 2000). However, these categories do not tell us *why* individuals have particular technological risk attitudes. But first, let us start by defining what we mean by a technological optimist or skeptic.

Technological optimism

Technological optimists believe in the liberating power of technology—modern medicine liberates us from disease and space exploration liberates us from Earth. This attitude is captured in the modernist[1] movement and is still a common attitude (e.g., Ridley 2010; Tierney 2010; Pinker 2018; Rosling, Rönnlund & Rosling 2018), particularly among Americans, with good reason. Over the past century, life expectancy has steadily increased worldwide with no signs of ending. Technological optimists have a basic faith in technology and require proof of harm to abandon any specific technology. An even more extreme

---

[1] Modernism, with its various manifestations in Western philosophy, architecture, art, literature, and so on, is embodied in Ezra Pound's command, 'Make it new!'

technophilia seems to be prevalent in Silicon Valley, where there is often effusive praise for all things related to the internet and related communications technology. The underlying assumption is these innovations have and will continue to fundamentally restructure human interactions to everyone's benefit. Some of the expected future benefits are as yet unimagined based on the observation that technology is frequently repurposed by its users in surprising ways. Technological optimists see future repurposing of technology as empowering and creative rather than potentially harmful. For technological optimists, even problems that are caused by technology have a technological solution (e.g., geo-engineering as a solution for global warming).

Technological skepticism

Technological skeptics reject the technology-as-panacea paradigm. This attitude is more closely represented in the postmodern era (along with small enduring enclaves of pre-modernists, such as the Amish) and is linked to some of the failings of modern industrialization. In particular, within the environmental movement, there has been an ongoing critique of modern Western society that includes a general aversion to technology (Leopold 1949; Carson 1962; Naess 1973). However, the roots of technological skepticism date back to at least the early 19th century as the Industrial Revolution was underway in Great Britain. The Luddite rebellion, a brief spasm of violence between 1811 and 1813, was a reaction to the social upheaval caused as the steam engine and power loom rapidly shifted the wool industry in central England from family cottage weavers to coal-powered mills run by a few adults and cheap child labor (Sale 1995). It is no coincidence Mary Shelley's *Frankenstein*, a seminal example of technological skepticism, was published in London in 1818. The full history of technological skepticism is too rich to cover here, but some notable efforts have been made to trace the thread of this philosophy to modern times (Fox 2002; Jones 2006).

Modern technological skeptics are more likely to question and critique the work of scientists and engineers. These criticisms are sometimes met with the reflexive scornful label of 'neo-Luddite,' but the attitude is not associated only with those who shun technology. The ranks of technological skeptics include engineers (e.g., Joy 2000) who recognize that our present-day society is privileged and powerful due to technology but that this technology hangs over our heads like a sword of Damocles.

Explaining the Differences

If we accept that there are differences in technological risk attitudes that can be roughly categorized, we invite questions regarding the nature and measurement of these difference. The following is a summary of some notable attempts, with varying degrees of success, to explain the differences in risk attitudes that are relevant to science and technology.

Cultural Theory

The cultural theory of risk proposes risk attitudes originate from cultural influences rather than from individual calculations of utility as proposed by the rational choice theory, which undergirds traditional risk analysis. That is, individuals filter information through cultural world views that determine their perception of risk. The theory categorizes four risk ideologies: laissez-faire *individualists*, social justice *egalitarians*, utilitarian *hierarchists*, and apathetic *fatalists* (Douglas & Wildavsky 1982). Individualists view human ingenuity as boundless and nature as robust. This roughly corresponds to technological optimism. Conversely, egalitarians, who roughly correspond to technological skeptics, view nature as fragile and have more precautionary views of technology. The hierarchists do not have strong attitudes regarding technology and are more likely to be ambivalent, but they do value authority, expertise, and the status quo (van Asselt & Rotmans 2002). Accordingly, hierarchists should be the most influenced by the risk statements made by authorities and experts. Lastly, the fatalists have little faith in the predictability of nature or humanity or in our ability to learn from past mistakes. Because fatalists doubt their capacity for control and self-determination, they often opt out of the policy-making process entirely. One implication of this classification scheme is risk analysts will tend to produce assessments that align with their own views and ignore the other cultural world views.

Despite its theoretical elegance, cultural theory has had limited predictive success (Marris, Langford & O'Riordan 1998; Sjöberg 1998). For example, cultural world view accounted for only three percent of the variance in surveys measuring the perception of risks and benefits of childhood vaccinations (Song 2014).

Psychometric measures

Explaining risk attitudes via psychological models, such as the risk-as-feeling concept (Slovic et al. 2004) has been more successful than cultural theory but still has limited explanatory power in empirical studies (Sjöberg 2000). Psychometric studies have been useful for explaining why public risk perception often deviates from calculated risks (e.g., Slovic 1987). They have also been used to directly investigate attitudes about technology and have found the acceptability of a technology is explained not only by its risk, but also by its perceived usefulness and whether it could be substituted with something else (Sjöberg 2002).

Additionally, technologies that tamper with nature and may have unknown effects are perceived to be riskiest.

However, even an expanded set of psychometric characteristics accounts for only a portion of the variance in technological risk attitudes, indicating the origin of these attitudes is still not well understood. For example, a survey of perceived risk from the chemical industry in South Korea found less than 10 percent of the variance was explained by cultural theory or psychometric measures, which was less than even basic demographic factors, such as education level or gender (Choi 2013).

Other theories

Other theories for why people embrace or reject technology have been proposed. For example, cultural cognition theory is a hybrid of cultural theory and psychometric models (Kahan et al. 2006, 2010, 2015). Cultural cognition recognizes a set of psychological mechanisms that allow individuals to preferentially select evidence that comports with the values of the cultural group with which they identify (Kahan, Jenkins-Smith & Braman 2011). In practical terms, this means evidence from experts that share the same values as their audience are viewed as more credible.[2] Likewise, information is more likely to be accepted if it can be presented in such a way as to be consistent with existing cultural values (Kahan 2010). Most people will reject new information if it cannot fit within their existing worldview or that of their social group.

This is consistent with psychologists' wry description of humans as 'cognitive misers' (Fiske & Taylor 1984). Most of us are not inclined to assess the veracity of every claim we encounter. But we cannot blame it all on innate laziness. To take the time to gather evidence and also acquire the expertise to evaluate the evidence can only be done on occasion—no one can be an expert about everything. Thus, most knowledge rests on a certain degree of faith in its source. This is where heuristics are used. We tend to believe claims made by people we personally trust or who have already been vetted by society—people with prestigious academic or corporate affiliations; people who are famous or are endorsed by famous people; or barring any other cues, people who are physically or emotionally appealing to us. This can cause trouble. For example, John Carreyrou's 2018 book, *Bad Blood*, recounts the hopeful rise and fraudulent fall of the blood testing company Theranos. The story serves as a cautionary tale for future technologists but also as a stark reminder of how the bandwagon effect

---

[2] This point is particularly useful for science communication. Many scientists have found more success through direct local engagement with the public rather than lecturing them via social media.

of our reputational system can catastrophically fail us when respected individuals are deceived by commanding personalities.

Other work in social psychology and sociology has also explored the interdependence of beliefs and how social groups can become intellectually closed off due to extreme skepticism of contradictory evidence provided by outsiders. However, the timing and interaction of influence between individuals and their social groups can be complex to model (Friedkin et al. 2016). Perhaps, as a result, cultural cognition theory has similar explanatory power to its parent theories. This hybridization may be the best of both worlds but still does not fully explain the source of risk attitudes.

Another theory comes from philosophers of technology. Framed in terms of trust, technological optimism and skepticism can be viewed as trusting or distrusting the reliability of technology as well as trusting or distrusting our ability to use technology appropriately (Ellul 1964; Heidegger 1977; Kiran & Verbeek 2010).

Sociologist Daniel Fox ascribed technological optimism to fatigue with the political process and considered it a misguided desire to resolve seemingly intractable social problems: 'The rejection of politics among intellectuals often takes the subtler form of what I call technocratic solutionism. Experts who practice solutionism insist that problems have technical solutions even if they are the result of conflicts about ideas, values and interests' (Fox 1995: 2). Understanding which problems require or are amenable to technological solutions is not always obvious. For example, conventional wisdom has usually attributed famines to food production failures—a technological problem. However, economist Amartya Sen found that famines in the past century occurred during times of adequate food production; the real culprits were hoarding and high prices brought about by bad governance and corruption—a political problem (Sen 1981).

Another idea not previously explored is the application of social psychologist Jonathan Haidt's theory of the ethical differences between liberals and conservatives (Haidt 2012). He argues there are six basic themes in moral thinking: prevention of harm, fairness, liberty, loyalty, authority, and sanctity. Furthermore, differences in political ideologies can be traced to which of these themes are emphasized. Specifically, American liberals judge the morality of actions primarily by two characteristics: does it cause harm and is it fair. Conservatives, on the other hand, appear more concerned with the additional criteria of loyalty, authority, and purity. For example, the purity concept is central to the 'wisdom of repugnance' championed by bioethicist Leon Kass, the chairman of President Bush's President's Council on Bioethics (Kass 1997).

This difference in conceptions of morality could partially explain why technological attitudes do not clearly align with political ideology. While both groups would be concerned about potential harm, it would be a priority for liberals. This might account for the liberal skepticism for technologies with potential environmental impacts but optimism for social media technology that is perceived to encourage equality. Likewise, conservatives might be more likely to accept military technologies that reinforce authority but reject medical technologies that offended their sense of sanctity.

This idea of the non-alignment of political ideology with technological acceptance is also captured in sociologist Steve Fuller's contention that the old left-wing/right-wing political ideologies are being supplanted with 'proactionary' (Fuller & Lipinska 2014) and 'precautionary' attitudes about science and technology that approximately correspond to technological optimism and skepticism, respectively. Dr. Fuller revives terms coined decades ago by futurist FM-2030 (born Fereidoun M. Esfandiary) that the left-right political distinction is becoming an up-down technological risk distinction—'up-wingers' embrace technological change as a source of opportunity and 'down-wingers' are more concerned about the risks of technological hubris. More importantly, the new ideological dichotomy is not merely a rotation of the old groups but a reordering. Up-wingers would be expected to come from the technocratic left and the libertarian right, while down-wingers would encompass environmentalists from the left and religious conservatives from the right. In terms of figureheads, think Elon Musk versus Pope Francis.

While these new labels provide some new insight into cultural trends, they remain just as coarse as the old ones and are no better at describing the complexity of individuals nor the source of their attitudes. For example, it is easy to imaging a person who is extremely supportive of medical research, cautiously optimistic about geoengineering as a temporary solution to climate change, but highly skeptical of pesticides, GMOs, and the value of social media.[3] How would we classify that person's technological ideology?

In the end, none of the individual theories presented here alone offer convincing explanations for the variation in attitudes about technological risk. The most persuasive trend is that multidimensional measures of risk perception tend to have more explanatory power than single factor explanations (Wilson, Zwickle & Walpole 2019)—that is, risk perception is complex. However, this collection of theories does give us a general idea of the range of factors at play in the formation of technological risk attitudes. Given our present inability to explain their origins, it may be more helpful to shift our focus from why these attitudes exist to the question of whether and how technological risk attitudes change over time.

How Do Technological Risk Attitudes Change?

How static are the ideas of technological optimism and skepticism? This question is important if we are to ascertain whether an event or additional

---

[3] Well, at least it is easy for me to imagine.

information, such as a research study or risk assessment, is likely to have a sub-stantial policy impact. Of course, influencing attitudes about technology is not as easy as simply presenting new information. Even risk communication meant to reduce concerns can have the unintended effect of raising fears about related risks among skeptics (Nakayachi 2013).

First, it appears cultural attitudes regarding technology change over time. The general public of the early and mid-20th century saw a steady stream of techno-logical wonders: radio, air conditioning, plastics, automobiles, airplanes, anti-biotics, television, nuclear power, and so on. The technological optimism fueled by this parade of inventions perhaps culminated with a man walking on the moon. Subsequent generations have failed to witness such large scale techno-logical spectacles, leading to a concern by some that society can no longer solve big technical challenges (Pontin 2012). Nonetheless, steady medical advances and the internet age have reignited technological optimism in certain segments of society. The futurist Ray Kurzweil has predicted artificial intelligence will exceed any human intelligence by 2030. His predictions of other computing milestones were considered unrealistic in the past, but his track record of suc-cess has made him mainstream in Silicon Valley such that he was hired as a director of engineering at Google in 2012.

Trends in technological optimism and skepticism can also be traced through science fiction literature. The romantic era of early science fiction, which encom-passed the second half of the 19th century, envisioned the utopian potential of technology. For example, Jules Vernes' submarine Nautilus in *20,000 Leagues Under the Sea* is a technological marvel used to explore and catalogue nature, aid the oppressed, and oppose militarism. Subsequently, the first few decades of the 20th century, dubbed science fiction's 'radium age' (Glenn 2012), saw a trend toward technological skepticism. Aldous Huxley's *Brave New World* (1932) is the epitome of the era. Like all literature, dystopian science fiction is a product of its time. Walter Miller's *A Canticle for Leibowitz* (1959) and Kurt Vonnegut's *Cat's Cradle* (1963) were inspired by the threat of Cold War nuclear annihilation. More recent trends have focused on biotechnology, such as Mar-garet Attwood's *Oryx and Crake* (2003).

While prevailing attitudes about technology have changed over time, they also vary geographically. One comparison of US and British attitudes regard-ing analog computing technology prior to World War II argues the US may have surpassed British efforts due to cultural differences regarding resistance to change and general enthusiasm for technological innovation (Bowles 1996). Likewise, emerging nations are considered more optimistic regarding the abil-ity of technology to solve problems than developed nations with strong envi-ronmental movements (Gruner 2008).

The variability of technology attitudes within an individual may be just as complex as temporal and geographical trends in society. Unlike the cultural theory of risk, the risk-as-feeling psychometric framework allows for vari-able attitudes about technology within a person. This agrees with our personal experiences where we may encounter individuals who are technological optimists in some fields of science, such as the potential for medical technology to improve people's lives, while being technological skeptics in others, such as blaming modern communications technology for loss of privacy.

It also seems reasonable that dramatic personal experience could substantially change an individual's technological risk attitude. The use of pharmaceuticals and medical devices, such as artificial joints, have become commonplace and can substantially improve quality of life. This not only makes technology more familiar, it can also greatly increase the technological enthusiasm of the person dependent on the technology. This appears to be the case for the technology theorist and transhumanism advocate Michael Chorost, whose hearing was restored by a cochlear implant (Chorost 2005, 2011), or Neil Harbisson, whose total color-blindness has been augmented with a sensor that converts not only color, but also IR and UV wavelengths to sound allowing him to detect things invisible to the human eye.

Conversely, technological optimism may evolve into skepticism when a job is made obsolete through technology. Travel agents, typists, toll booth attendants, telephone operators, video store clerks, and photo lab employees are just a few of the many careers that quickly appeared, seemed as though they would always exist, and then just as suddenly disappeared due to disruptive technological change. Technology-induced obsolescence of this nature is now even affecting many of the skilled professions (Rifkin 1995; Rotman 2013).

Principle of Dangerous Science

So far we have discussed the origins and flexibility of technological risk attitudes. Despite the lack of a comprehensive theory, it appears these attitudes are influenced by a variety of factors that include culture, feelings, and personal circumstances. They also appear to be malleable over time at both the individual and societal level. Do these observations have any implications for how science and technology policy decisions are made? Based on the complexity of technological risk attitudes, can we make any general statements?

First, let us start with the additional observation that few lines of research or emerging technologies have been banned or abandoned in the past for reasons unrelated to science or practicality. Many controversial lines of research and technologies, such as the practice of lobotomies, have been abandoned for lack of scientific value and better alternatives.[4] However, the list of scientifically valid research that was stopped for moral reasons or social concerns is relatively short and consists primarily of weapons technology; internationally

---

[4] The inventors of the procedure were awarded the 1949 Nobel Prize in Physiology or Medicine, which reflects rather poorly on our ability to predict the long-term value of research.

banned research includes biological and chemical weapons research as well as weather modification for military purposes (the 1978 Environmental Modification Convention). Likewise, the list of highly restricted research—for example, embryonic stem cell research in the US—is also relatively small. Despite substantial ethical concerns or public opposition, a wide range of controversial and non-essential research activities are currently legal in most places.

In general, the number of technologies and lines of research banned for ethical or risk perception reasons is small enough to suggest a general principle of dangerous science. The principle is simple: research is not banned solely for moral reasons. No line of research or technology will be forbidden until it has been scientifically discredited or deemed impractical or a better alternative has been accepted in its place. The implications of such a principle are substantial if, as sociologist John Evens (2018) argues, the vast majority of conflicts between religion and science are moral debates, not disagreements over knowledge.

A precursor to this principle is the premise if something can be done that appears to have some intellectual or economic value, someone will view it as a valid idea and proceed before anyone else beats them to it. This then leads to the principle that posits we generally do not stop research just because it may be unwise, potentially harmful, or otherwise ethically dangerous. The result is dangerous science generally moves forward until something eliminates the controversy. The controversy can be eliminated in one of several ways: new information or extensive experience reduces the perceived danger, cultural acceptance decreases opposition, or an alternative technology eliminates the need for the research.

Underlying this principle is a pervasive attitude of inevitability surrounding technology in modern society. The sense technology controls us as much as we control technology is described by social scientists as 'technological determinism' (Bimber 1994). Popular books on the philosophy of technology (e.g., Arthur 2009; Kelly 2010) adopt this mindset when they describe the evolution of technology and occasionally use the term in the biological sense. By using multiple meanings of evolution, these technologists reveal their mental image of technology as independent and alive and perhaps uncontrollable. Scholars of the field of science and technology studies have long argued technology, by its very anthropogenic nature, is embedded with human values (e.g., Jasanoff 2016). Technology requires intentional and unintentional value-laden decisions in its creation, and the more complex the technology, the longer the string of decisions. The result is technology is quite controllable *if* we make thoughtful and compelling choices to do so. Thus, technological inevitability may really be a method for remaining uncritical about the results of human endeavors. In a 1954 hearing, J. Robert Oppenheimer, the lead physicist of the Manhattan Project, admitted an Achilles heel of scientists left to their own devices: 'When you see something that is technically sweet, you go ahead and do it and you argue about what to do about it only after you have had your technical success. That is the way it was with the atomic bomb.' Perhaps the primary force behind technological inevitability is merely the pull of an intellectual challenge heedless of the consequences. Compounding the problem is the centrality of technology to our lives, which exerts a powerful force on what activities we believe are possible, our behaviors, and our expectations. This can support the illusion the present path of innovation is the only option.

Continuing with this biological metaphor, technological optimists think of technology as benignly natural and inevitable, while technological skeptics see it as aggressively virus-like and relentless. If the history of technological progress gives the appearance of supporting the concept of inevitability, technological skeptics see this inexorable pull as more ominous: 'we are free at the first step but slaves at the second and all further ones' (Jonas 1984: 32).

The emphasis on describing the trajectory of technological progress also has implications for policy making. With an assumption of inevitability, technological discussions tend toward predicting what will come next rather than discussing what *should* come next. While there are academic communities working in specific subfields of technology ethics, such as bioethics, the broader technology ethics community is surprisingly small considering the criticality of technology in modern society.[5] Likewise, innovation literature is primarily focused on how to encourage innovation rather than normative discussions of where innovation should lead. The idea of 'responsible innovation' has only started to gain traction in this century (Guston et al. 2014). If we fail to actively direct science research and technological innovation, what at first glance feels like technological determinism is actually 'technological somnambulism' (Winner 1986).

The principle of dangerous science appears to be particularly true for emerging technologies, which are noted for rapid change. When the time between basic research and usable technology is very short, the process of societal reaction, evaluation, and public policy formulation often lags the rate of technological progress. Likewise, because the research and its technological applications can be nearly concurrent, available empirical evidence may be of limited value in estimating the future trajectory of the technology (Chameau, Ballhaus & Lin 2014). In the absence of convincing data or tractable theoretical frameworks, policymakers tend to favor cautious permissiveness until more compelling evidence is available. There is a pervasive fear of limiting technology development, especially in the face of marketplace competition, international military rivalry, industrial anti-regulatory lobbyists, and science proponents raising alarms of innovation deficits. No one wants to be seen as hampering scientific progress because of speculation or slippery slope arguments (even if they are not unwarranted concerns). Permissiveness, at least in the US, is often a defensible default

---

[5] There isn't even agreement on whether there should be more. While some are alarmed by the paucity of bioethicists working in the field of dual-use biotechnology (Rappert & Selgelid 2013), others argue the professionalization of bioethics removes important discussions from public debate (Cameron & Caplan 2009).

position because regulatory agencies are frequently unprepared and lack jurisdiction until public sentiment pushes policymakers to react.

Here, we finally come back to the idea of competing technological risk attitudes that exacerbate technological inevitability in policymaking. Technological optimism and skepticism function as opposing ideologies, which tend to limit the use of appeals to common values as compelling reasons for technology regulation. This creates a society where there is plenty of ethical assessment of science and technology going on—institutional review boards, national and international science ethics committees, governmental regulatory agencies, et cetera—yet the collective oversight tends to narrowly define its mission toward preventing only the most egregious ethical lapses. The result is a sympathetic ethical review process that gets caught by surprise when occasional research causes public alarm and the obligatory *post hoc* regulatory scramble.

The following four examples further explore the principle of dangerous science within the context of contemporary emerging technologies. The first three serve as examples of the principle in action, while the fourth is a potential counterexample.

The Principle in Action

Womb transplants

While organ donation is a well-established lifesaving procedure, non-essential organ transplants are more controversial. In 2014, the transplantation of a uterus from a post-menopausal woman to a woman of child-bearing age without a functional uterus resulted in the birth of a premature, but otherwise healthy, child (Brännström et al. 2014). Several bioethicists noted the procedure was not medically necessary and was rather dangerous. It required transplanting not only the uterus but also much of the surrounding uterine vasculature during a lengthy surgery. The recipient then had to live with the risks associated with immunosuppressant treatments and the potential of transplant rejection (Farrell & Falcone 2015). Furthermore, uterine transplants are intended to be temporary, and a second surgery is required to remove the donated uterus so the patient can discontinue immunosuppressant therapy after successful childbirth. While the surgeons involved appeared to be deeply concerned for the safety and mental well-being of their patients, who were desperate to experience pregnancy, there was also the unsavory appearance of a race for personal scientific glory to be the first to successfully perform the procedure (Orange 2014).

Fundamentally, the question of this research comes down to whether one believes it is acceptable to perform risky non-essential surgery for personal, cultural, or religious reasons. While medically-unnecessary procedures are performed on a regular basis, the level of risk to the patient in this case is substantial. Furthermore, from the perspective of the child, transplant pregnancies have much higher health risks. Bioethicist Julian Savulescu has argued for a principle of 'procreative beneficence' that requires parents to maximize the health of their future children. From this perspective, having a child via uterine transplant should be avoided because it is a higher risk to the child's health compared to alternatives (Daar & Klipstein 2016). Nonetheless, proponents argue the work passes the test of clinical equipoise—the potential benefits exceed the risk of unintended harm to the mother, child, and donor (Testa, Koon & Johannesson 2017).

The principle of dangerous science suggests this line of research will continue, especially once it was proven successful. The first success outside of Sweden occurred in Texas in November 2017, which further encourages medical professionals elsewhere. However, given the considerable expense, the high failure rate, and the availability of alternatives, such as surrogacy or adoption, it is likely the procedure will remain relatively uncommon. It is also interesting to note Swedish surgeons are world leaders in uterine transplants—eight successful childbirths had occurred by the end of 2017. Their success is partially due to a lack of alternatives because Sweden discourages the use of surrogate mothers, which is, ironically, deemed unethical.[6]

While the procedure is unlikely to be banned for ethical concerns, womb transplants may be eventually abandoned due to impracticality. The true end of the procedure will likely come when scientists are able to grow a new functional uterus from a woman's own stem cells—a technically (and ethically) superior alternative.

Human genome editing

While gene manipulation using recombinant DNA techniques is decades old technology, new techniques and recent research have led to proposals of germline editing (making genetic changes in embryos that are inheritable). This has raised concerns over potential multi-generational harm as well as a public backlash against meddling with nature. In early 2015, the British parliament approved three-person *in vitro* fertilization (IVF), where a third person provides mitochondrial DNA to correct mitochondrial diseases. Some see this as a reasonable first step but larger scale genome editing must await confirmation

---

[6] A growing number of governments agree that surrogacy is exploitation—particularly when poor foreign women are paid to rent their wombs. Surrogacy has been banned in some jurisdictions and restricted to altruistic or family surrogacy in others. At first glance, this appears to be a potential exception to the principle of controversial research. However, the problem is not with the science itself, which is commonly used in fertility clinics, but with the legal contract of surrogacy. Or maybe this is splitting hairs.

it is safe (Baltimore et al. 2015). Although a 2016 US budget bill banned FDA funding for human embryo heritable genetic modification, the first successful mitochondrial replacement therapy by three-person IVF occurred in New York in 2016, resulting in the birth of a baby boy (Zhang et al. 2017). Because mitochondria are inherited from mothers, the technique does not result in heritable genetic modification in male children, although there appear to be rare cases of mitochondrial DNA inheritance from fathers (McWilliams & Suomalainen 2019). However, when the New York case was published in an academic journal, there were concerns regarding questionable institutional review board approval, the level of informed consent, and the potential health effects of small amounts of diseased mitochondria transferring to the child (Alikani et al. 2017). Even scientists with substantial financial stakes in the new gene editing technology have argued parents should use genetic testing and screening options instead, reserving the new technology for the treatment of disease in somatic (non-germline) cells (Lundberg & Novak 2015).

Germline editing has been compared to the initial concerns regarding IVF, which was initially controversial but rapidly became a widely used medical technology. This may be an appropriate comparison. In 2018, a clinic in Kiev, Ukraine, acknowledged publicly it was offering three-parent IVF as a treatment for infertility. This new application of mitochondrial replacement had already resulted in the birth of four children, one of whom was female and may presumably pass on her genetic modification to her children.

Chinese scientists were the first to attempt germline editing with non-viable embryos obtained from fertility clinics to test the potential for curing a fatal genetic blood disorder (Liang et al. 2015). While the initial study had a low success rate, the technology rapidly improved (Liang et al. 2017). Given the vast number of genetic diseases that appear amenable to treatment by genome-editing (Ma et al. 2017), cautious support for this research appears to be growing. A 2017 report from the Committee on Human Gene Editing of the National Academies of Science and Medicine shifted the general expectation from total prohibition to limited approval (National Academies 2017b). Likewise, surveys seem to indicate, within a few short years, the majority of the public has already accepted the idea of inheritable genome edits to prevent disease (Scheufele et al. 2017).

Some critics have called for a complete ban, claiming germline editing constitutes 'a return to the agenda of eugenics: the positive selection of "good" versions of the human genome and the weeding out of "bad" versions, not just for the health of an individual, but for the future of the species' (Pollack 2015). Others are concerned germline editing will cause the public to restrict all gene editing activities. However, history and the principle of dangerous science suggests otherwise.

There have been few calls in the US for moratoria in the biological sciences: recombinant DNA research in 1974, human reproductive cloning in 1997, and influenza gain-of-function research in 2012. Of these three moratoria over the past 40 years, the recombinant DNA moratorium was lifted within a year (Berg & Singer 1995). The third moratorium on gain-of-function research was lifted in 2017 as detailed in the first chapter. The human cloning moratorium is still unofficially in effect for federally-funded research; however, human cloning was never actually banned in the US despite legislative attempts in 1997, 2001, 2003, and 2005. The moratorium was originally inspired by a spate of large animal cloning successes (for example, the birth of Dolly the sheep in 1996). However, once it was found defect rates were high in the cloned animals, commercial interest in cloning large animals diminished. Nevertheless, the research continued, and within 20 years a Korean laboratory began marketing pet dog cloning services. Even with a low success rate that requires many donor and surrogate dogs, as well as a $100,000 price tag, the ethically questionable lucrative niche market supports the principle of dangerous science.

In early 2018, the first successful cloning of primates was announced in China. The intent of cloning macaques was to improve medical testing through the use of genetically identical animals. Successful cloning of a primate suggests the hurdles to human cloning are negligible. In the end, the US ban on human cloning did not stop research from walking right up to the forbidden line. If a new biotechnology has a use with tempting benefits, moral concerns provide a weak barrier.

In late 2018, Chinese researcher He Jiankui announced he had edited the genome of recently born twin girls to prevent expression of the white blood cell receptor protein CCR5 with the intent of preventing HIV infection. The edit emulates a natural variant, CCR5-Δ32, which provides HIV-1 resistance to some European populations. It is posited this variant confers resistance to some types of viruses but may also increase susceptibility to other viruses, such as West Nile and influenza. Given the lack of understanding of the implications of this particular edit, and CRISPR[7]-mediated germline editing in general, the work was considered extremely premature (Ledford 2019). Along with widespread condemnation by the international scientific community, an international germline editing moratorium was proposed (Lander et al. 2019). Dr. He was terminated from his university position, a government investigation was launched against him, and the possibility of criminal charges were discussed. The work even led to an investigation of a Stanford University post-doc advisor to Dr. He who, despite being cleared of any wrongdoing, became the subject of an unflattering *New York Times* article. However, even after the intensity and breadth of all this criticism, a Russian scientist subsequently announced interest in performing a similar germline editing procedure (Cyranoski 2019).

---

[7] Since first proposed in 2012, precision gene-editing tools based on CRISPR (clustered regularly interspaced short palindromic repeats) DNA found in prokaryotes have greatly advanced the genetic engineering abilities of scientists.

Synthetic biology

Regardless of whether one considers it a new field or an old one with a new name, synthetic biology has experienced a similar level of promise and controversy. Although rather amorphous and evolving, the field entails purposeful genetic manipulation coupled with the engineering philosophy that creation is more instructive than observation (Roosth 2017). The field has two complementary approaches: editing existing genomes with tools such as CRISPR-Cas9 or the more tedious process of experimentally building 'minimal genomes' (Hutchison et al. 2016) from scratch that contain only essential and functional genetic material. The engineering mindset of synthetic biology also emphasizes problem solving. The techniques have already been applied to the production of biofuels, pharmaceuticals, and food additives.

Applications started with viruses and bacteria before moving on to simple eukaryotes. For example, the Synthetic Yeast Genome Project (Sc2.0) has a goal to re-engineer and simplify the 12 million base pairs of DNA in the 16 chromosomes of baker's yeast, *Saccharomyces cerevisiae* (Richardson et al. 2017) so it may be used as a miniature factory on which to base a myriad of other functions. The expectation is that plants and animals are the next step.

The field brings to reality some seemingly fantastical possibilities, including the creation of organisms resistant to all viral infections (Lajoie et al. 2013)—a useful trait in bioreactors but also a major competitive advantage in nature—or even the creation of new forms of life (Malyshev et al. 2014). With it comes immense commercial potential (Gronvall 2015) as well as a host of new social and ethical considerations (Cho et al. 1999).

Expert opinions on synthetic biology vary widely. Technological optimists claim detailed modification of genetic sequences will result in decreasing unintended consequences. Likewise, synthetic species that cannot mate with wild species should, in theory, greatly reduce the likelihood of unintended gene transfers or uncontrolled releases. Meanwhile, technological skeptics fear unpredictable emergent properties, scientific hubris, and expanding biosecurity threats. With the potential for substantial benefits and harm, questions have been raised regarding whether the current regulatory systems and avenues of risk assessment are sufficient (Lowrie & Tait 2010; Kelle 2013; Drinkwater et al. 2014; National Academies 2018a). However, despite the reoccurring fears of scientists playing God or references to *Frankenstein*, it seems, much like with the human cloning debates, the primary concern of bioengineering is maintaining the specialness of human life (van den Belt 2009). As long as that line is not crossed, the synthetic biologists will probably be allowed to tinker on.

The many uses already imagined for synthetic biology in just the fields of medicine and manufacturing alone suggest a future that could be much healthier, cleaner, and more energy-efficient. Of course, like all powerful science, one person's commercial imperative is another person's irresponsible pursuit. One emblematic example is the successful bioengineering of yeast to convert sugar into morphine (DeLoache et al. 2015; Fossati et al. 2015; Galanie et al. 2015). The original work could not generate commercially viable production levels, but to critics the research suggested home-brewed heroin and all its associated social problems were inevitable (Oye, Lawson & Bubela 2015). Opioid bioengineering would appear to be a classic example of the principle of dangerous science in action. The work proceeded in the midst of a raging opioid addiction epidemic in the US, and the general timing of announcements from competing research teams suggested the artificial urgency of a race for personal glory. Given the plentiful supply of raw materials and ease of manufacturing, there was no public health call for more accessible opioids (McNeil Jr 2015). Interviews with the researchers suggested they believed their work would help address existing pain-management crises in less industrialized countries. However, any pain medication shortages in those countries have been primarily caused by policy decisions rather than unaffordable medication.

Despite the obvious potential for misuse, the research was funded because opiates derived from yeast rather than opium poppies could potentially reduce opioid production costs. Benefits to areas, such as Afghanistan, where opium poppy production provides funding for military conflicts also served as further theoretical justification. On top of this, the researchers created a biotech start-up, Antheia, to ramp up production efficiency and, perhaps to mollify their critics, suggested the potential for creating less addictive opioids in the future. Besides private investors, Antheia's work was funded by the NSF, NIH, and the National Institute on Drug Abuse. Economic benefits aside, questions arise regarding the likelihood an improved process can be kept out of the illegal opioid market. Likewise, substituting poppy-derived opioids only impacts legal farming in India, Turkey, and Australia. The illegal opioid market fueled by Afghanistan would be unaffected unless the illegal market also converted to the new process. The end result is a formidable regulatory and public health challenge if the technology succeeds.

Autonomous weapons

Technologies developed specifically for military purposes, such as bioweapons research, appear to serve as the counterexample to the principle of dangerous science. Research on specific classes of weapons are one of the few types of research banned in the past.[8] However, it could be argued weapons research

---

[8] Efforts can be legislative or legal actions. For example, efforts in the US to sell plans to print untraceable 3D plastic guns has been thwarted by multiple court injunctions. However, the primary obstacle remains the technical skill and expensive equipment still required to create a functional weapon.

bans are merely a special case in the sense the beneficial uses of weapons (e.g., deterrence) are limited and can often be accomplished with less controversial alternative technologies. It should also be noted some weapons research has continued in secret even after the research was banned, such as the bioweapons program in the former Soviet Union (Leitenberg, Zilinskas & Kuhn 2012; Sahl et al. 2016), thus conforming to the principle of dangerous science.[9]

One contemporary example is the field of autonomous weapons. This century has seen an explosion in the number of drones equipped with lethal weapons. Robotic weapons in various stages of development now include the option of fully autonomous operation (Chameau, Ballhaus & Lin 2014). Proponents argue autonomous weapons can minimize civilian casualties, while critics are concerned these systems will lower the social cost of going to war and will eventually be used universally by both police forces and terrorists (Russell 2015). There have been efforts to ban autonomous weapons through the Convention on Certain Conventional Weapons led by computer scientist Noel Sharkey, who chairs the nongovernmental International Committee for Robot Arms Control. Its mission statement includes the reasonable premise 'machines should not be allowed to make the decision to kill people.' Another effort, the evocatively named Campaign to Stop Killer Robots, was launched in 2013 to raise public awareness about lethal autonomous weapons and to persuade prominent scientists and technologists to publicly denounce such systems.

Autonomous weapons systems may have the best chance of being banned solely on moral grounds for two reasons. First, a history of international weapons treaties—the 1968 Non-Proliferation Treaty, the 1972 Biological Weapons Convention, and the 1993 Chemical Weapons Convention—suggest weapons technologies create a unique broad consensus regarding their risk-benefit assessment. Second, the 1997 Ottawa Treaty, which bans landmines, creates a specific precedent for banning (albeit simple) autonomous weapons.

Whether autonomous weapons are actually banned remains to be seen. Recent books on the subject tend toward pessimistic inevitability while still offering warnings and hope that humans will decide otherwise (Singer 2009; Scharre 2018). However, prospects are dimming as technology advances and the extraordinary becomes familiar. Countries leading the rapid development of this technology—including the US, the United Kingdom, and Israel—have argued against any new restrictions. In 2018, a research center for the Convergence of National Defense and Artificial Intelligence was opened in South

---

[9] Likewise, the US Defense Advanced Research Projects Agency has funded research to create food crop horizontal gene transfer systems (as opposed to vertical transfer through inheritance) that use gene-editing viruses delivered by insects. The purported intent is to protect crops within a growing season rather than waiting to protect subsequent crops. However, the simplest use of this problematic technology is as a bioweapon (Reeves et al. 2018).

Korea despite protests from computer scientists in other countries suggesting that work by a respected research university was incrementally moving autonomous weapons into the realm of normality. In a 2016 *Agence France-Presse* interview, computer scientist Stuart Russell stated, 'I am against robots for ethical reasons but I do not believe ethical arguments will win the day. I believe strategic arguments will win the day.'

Implications

So how might the principle of dangerous science guide debates over emerging technologies? One observation is that technological optimists have history on their side. They need not overhype new technology because it is unlikely to be banned or heavily regulated until there is overwhelming evidence of harm. On the contrary, more voice could be given to the technological skeptics without fear of stifling innovation.

For example, one particularly energetic defense of new gene editing techniques, such as CRISPR-Cas9, called for bioethicists to 'Get out of the way' (Pinker 2015). However, such advice is misplaced. Institutional review boards are not the main culprit delaying life-saving technology. Increasing research funding would improve far more lives than streamlining bioethics reviews. Of course, criticizing the obscure work of review boards is easier than simply asking taxpayers for more money. History suggests gene editing techniques are far too useful to be limited by more than cursory regulation. Once these powerful techniques becomes ubiquitous, a more realistic concern is they will be difficult to control and society will have to depend upon the moral wisdom of bioengineers.

Coming back to the avian influenza research discussed in the first chapter, are there some further insights available from the discussion of technological risk attitudes? Discrepancies in biotechnology risk estimates have been attributed to two competing narratives on technology development: biotech *revolution* and biotech *evolution* (Vogel 2013). The revolution narrative is a dystopian form of technological determinism that assumes biotechnology development experiences rapid inevitable progress (Smith & Marx 1994). This view predominates the biosecurity community and has heavily influenced US biosecurity policy (Wright 2006). The evolution narrative is based on a sociotechnical model of technology development and takes the view that biotechnology is built on slow and incremental innovation (Jasanoff 2004; Nightingale & Martin 2004). The revolution narrative roughly equates to the skeptical technological risk attitude, while the evolution narrative is a more benign and optimistic technological risk attitude. The biotech evolution narrative also forms the basis for incrementalism, an argument frequently used by proponents of research freedom. Each published paper is usually a small addition to the corpus of scientific knowledge. Thus, if previous papers were not restricted, then why limit the next one?

This is why regulating entire lines of research rather than individual projects can be a more effective approach (Rappert 2014).

At the December 2014 NRC symposium on H5N1 research, biodefense policy expert Gregory Koblentz, discussed the propensity of risk attitudes to dominate risk assessments where there are sparse data. In the context of biosecurity risks, the optimists believe bioterrorism risk is exaggerated because few past terrorist attacks used bioweapons, terrorists tend to use more readily available weapons, and the technical obstacles are significant. Conversely, pessimists believe bioterrorism risk is understated because past uses of bioweapons show terrorists are innovative, terrorist acts have become increasingly lethal over time, terrorist ideologies embrace mass casualties, and technical obstacles are decreasing with time. The parallels of these opposing views to the broader categories of technological optimism and skepticism are clear. The dichotomy of views is partially due to the limited historical record of biological weapons, which leaves a great deal of room for interpretation (Boddie et al. 2015; Carus 2015, 2017). More importantly, these opposing views result in very different and controversial risk management strategies.

To summarize, in the absence of convincing evidence, technological risk attitudes often guide decision making. For the purposes of policy formulation, we can roughly separate these attitudes into technological optimism and skepticism which acknowledges reasonable people can reach opposing conclusions based on the same available data. These labels are pragmatic descriptions in the absence of a theory that satisfactorily explains the origins of technological risk attitudes. An observation from a review of these attitudes is it is unclear how to change them. They are influenced by many factors, and any single assessment is unlikely to change risk attitudes and thereby resolve a controversial policy debate. History also suggests research is rarely abandoned for moral reasons. If technological skeptics want to be more effective in making their case for caution, they need to form pragmatic arguments for their position and propose alternative solutions that address existing technological needs.

Managing Dangerous Science

So how might knowledge of the inherent subjectivities of risk-benefit assessment and the principle of dangerous science guide the management of science and technology? Before recommending improvements to the assessment and management of dangerous science, it is helpful to first review and critique some ways research has been managed in the past. One reason new approaches are needed is the changing nature of modern science.

The Changing Face of Technological Risk

Much of the life science research discussed in this book represents a relatively new category of research—technologies that are both accessible *and* powerful. Whereas chemical technology was the focus of dual-use concerns in the first half of the 20th century and nuclear technology in the last half, biotechnology is the most significant challenge to science policy in the 21st century. The primary reason is many hurdles in the biosciences are now more theoretical than practical—material resources are rarely the limiting factor. Laboratories around the world, including an increasing number of small unaffiliated labs, can now create or order with relative ease what was only possible at a handful of state-of-the-art facilities a few years before (Suk et al. 2011; Adleman 2014).

A personal anecdote demonstrates my point. In 2018, I attended a local elementary school science fair in a New York City suburb. Among the typical experiments displaying molding breads and growing plants, there sat a fifth-grade experiment titled 'Biohack'. With the assistance of a parent (a trained biologist), the student had used a mail-order CRISPR-Cas9 kit (Sneed 2017) to alter a non-pathogenic *E. coli* bacteria to become resistant to the antibiotic streptomycin. While this relatively harmless gain-of-function experiment was essentially a recipe for amateurs to follow, the fact a talented fifth-grader was

How to cite this book chapter:
Rozell, D. J. 2020. *Dangerous Science: Science Policy and Risk Analysis for Scientists and Engineers.* Pp. 77–102. London: Ubiquity Press. DOI: https://doi.org/10.5334/bci.e. License: CC-BY 4.0

able to successfully perform detailed genetic engineering at home suggests the potential pervasive power of this technology. Unfortunately, in 2017, authorities in Germany—where biological research is more strictly regulated—discovered a mail-order CRISPR kit from the US contained multidrug-resistant pathogenic *E. coli*. The risk to healthy users was deemed to be low, but the product was banned in Germany. Any regulators with qualms about the DIY-bio movement now had evidence to support their misgivings.

The critical component in many of today's emerging technologies is knowledge. Once the information is publicly available, the ability to create both immensely beneficial and harmful biotechnology becomes almost ubiquitous. The biggest remaining barrier is bridging the gap in tacit knowledge—the assumed professional knowledge, ignored or hidden details, and essential laboratory skills that are not recorded in the academic literature. However, even these hurdles are decreasing as modern communication lowers the cost of detailed knowledge transfer and as increasing numbers of experienced biotechnologists migrate from lab to lab (Engel-Glatter 2013; Revill & Jefferson 2013). Furthermore, many of the highly skilled and technical steps are being removed through automation. Together, these factors point toward a field that is rapidly deskilling (Roosth 2017), with all its associated implications for the careers of biologists and public biosafety. Managing potentially harmful biotechnology requires a fundamental shift in thinking from nuclear weapons nonproliferation policy, which historically focused on controlling materials as much as knowledge (Moodie 2012).

Furthermore, managing biotechnology materials are far more difficult than nuclear materials because they reproduce and are not easily detected. For example, a transgenic orange petunia developed in Germany in 1987, but never sold to the public, was found in 2015 to be commercially available unbeknownst to regulators and breeders. The result was a 2017 request to destroy an untold number of petunia varieties in the US because they contained DNA from cauliflower mosaic virus, which is listed as a plant pest. While the physical danger posed by these flowers is minimal, the regulatory fallibility they represent is not.

One outcome of the new power of biotechnology is the biological science community is now finding it increasingly difficult to self-regulate. Instead of just worrying about harming patients, institutional review boards are now confronted with the reality of 'bystander risk' (Shah et al. 2018) and unknown implications to society. The H5N1 research example hints at the difficulty of reactively managing dangerous science. After research has been conducted and technologies developed, responses tend to be *ad hoc* and difficult to implement. This raises important questions of how to proactively manage technologies that have a high potential for negative consequences but are difficult to govern because the technologies are already widespread and knowledge, rather than materials, is the limiting factor (Miller et al. 2009). Suggested approaches for these cases focus on informal governance measures that strengthen the international culture of responsible research (National Academies 2018b) and rely on moral persuasion methods, such as codes of conduct, educational outreach, transparency, and whistle-blowing support (Tucker 2012). One clever suggestion has been to use the patent system as a tool for regulating gene editing while in its early stages (Parthasarathy 2018), but this works best for large organizations; the DIY culture is less amenable to formal methods of oversight. Harnessing the power of social norms may be one of the few means of guiding the DIY-bio community (Nyborg et al. 2016). However, this requires a community where nearly everyone is doing safe experiments, and they consistently ostracize those who are perceived to be reckless. It is also helpful if trained scientists are actively engaged with citizen scientists to ensure they are using the safest methods available.

Community self-monitoring through actions, such as anonymous reporting systems, can substantially reduce obvious avoidable accidents. However, they appear to be an inadequate response to threats of intentional misuse because bad actors often attempt to hide their work. Another avenue being pursued is the use of AI automated screening tools for ordered DNA sequences. Again, this assumes bioterrorists, rather than working in the shadows, will be using commercial DNA synthesis to speed up their misdeeds.

While there is little consensus on how likely the actual bioterrorism threat may be, it is difficult to monitor and control bioterrorists within a largely self-regulated community. One cannot expect virtuous behavior from scientists and engineers who, by definition, have less than virtuous intentions. One example is the deadly 2001 anthrax letters that appeared to have originated from an anthrax biodefense research scientist who worked at a highly monitored US Army research facility on a military base (DOJ 2010). The lack of direct evidence, even after the prime suspect's suicide, suggests our ability to counter, or even detect, the rare 'mad scientist' is insufficient. Furthermore, just because the emphasis in the H5N1 debate eventually shifted to accidental release, there was no reason to dismiss the real threat of H5N1 bioterrorism (Inglesby & Relman 2016). Given the few viable options for managing low-governability technologies, involuntary monitoring of research activities by government intelligence agencies is a more likely future scenario.

Common risk management strategies

There are a few general criteria traditionally used to manage risk. First, one can attempt to maximize utility—seek maximum benefits for the minimum risk—which naturally leads to assessment by risk-benefit analysis. Second, one can attempt to solely minimize risk (zero risk being ideal) which leads to adopting a precautionary approach to technology management. Precautionary technology bans are a more common choice for extreme risks with deep incertitude. However, precaution does not require complete bans; it can also take the form of cautiously deliberative support (Kaebnick et al. 2016). Lastly, one can take a pragmatic technology-based risk minimization approach. This is often used in pollution control, where regulations call for 'as low as reasonably achievable' (ALARA) or 'best available control technology' (BACT). The first and third approaches are science-based strategies most useful when data are plentiful. However, these approaches are of less value for emerging dangerous science. Precautionary management is more easily applied to these situations, but the fear is it stifles innovation and progress.

Critique of traditional risk management

Predicting the future is a popular pastime with generally limited success. Unanticipated events with no historical precedent occur with surprising frequency (Taleb, Goldstein & Spitznagel 2009). Yet we tend to be overconfident in our assessment of risk and good at rationalizing inadequate risk management after the fact (Paté-Cornell & Cox 2014). Some of the excuses for poor risk management include claiming uncontrollable forces (acts of God), unimaginable events (black swans), rare confluences of events (perfect storms), lack of precedent, excusable ignorance, conformance to standard practice, someone else's responsibility, lack of legal liability, or even operator error. *Post hoc* risk reduction is so riddled with biases that realistic reflection and improvement is deceptively hard. The difficulty warrants adopting a more cautious risk management attitude that emphasizes the incompleteness of formal risk assessment and promotes a general skepticism toward quantitatively bounding the certainty of our knowledge (Aven & Krohn 2014). While it is important to acknowledge the limits of human prediction, superficial or non-existent risk assessment is much worse and promotes a slide into technological fatalism.

Previous efforts to improve the objectivity of science and technology assessments for policy decisions have proposed variations on the 'science court' (Kantrowitz 1967; Field Jr. 1994). The idea is to have science experts argue their position in front of a disinterested scientist from another field who would act as a judge or mediator. This would separate the judge and advocate positions as is common in the US judicial system. The proposal is an attempt to be proactive considering research controversies will eventual find their way to the legal system if not adequately addressed elsewhere.

Coming back to the case study in the first chapter, a series of newsworthy biosafety lapses at CDC laboratories in the summer of 2014 were key events leading to the moratorium on H5N1 virus gain-of-function research. These breakdowns in safety at a well-respected institution raised alarm among the public and policymakers. The standard assurances research was being conducted safely were now questionable. The result was a precautionary backlash typical of a 'popular technological assessment' (Jasanoff 2003). Such public responses often appear to ignore scientific evidence. However, they are not simply due to scientific illiteracy or incompetence,[1] but rather a reaction to sophisticated analyses that lack credibility (NRC 2015). Calls for precautionary bans were not restricted to regulators and the public. Biosecurity experts also saw them as a reasonable response to the threat of avian influenza research gone awry (Klotz & Sylvester 2012).

Because the research was not banned outright, one cynical interpretation of the entire debate was the various moratoria were politically necessary until public attention turned elsewhere and the final risk-benefit assessment was a fig leaf to justify business as usual. It was noteworthy the formal deliberations were not affected by a 2015 avian influenza outbreak in the US—the largest in decades. While the outbreak only affected poultry, the geographic immediacy and economic losses should have refocused public attention on the debate, but its impact was minimal. Given the underlying culture of technological inevitability prevalent in the US, perhaps the outcome is unsurprising. Nonetheless, the resulting minimally restrictive research policy for avian influenza was still a substantial shift from only a decade ago when both the NRC and NSABB broadly supported essentially complete self-regulation in the biosciences— a sentiment still popular in the EU where the European Academies Science Advisory Council still prefers to defer to the professionalism of the scientific community (Fears & ter Meulen 2016).

It is also important to note these moratoria only applied to research funded by the US government. It would be optimistic to assume US policy on potentially pandemic pathogen research carries as much weight as the Recombinant DNA Advisory Committee, which issues guidelines that are mandatory only for NIH-funded research but have become widely accepted (Resnik 2010). While the influence of federal funding within the global research community is still substantial, it is shrinking. It is not inevitable that even closely aligned communities, such as the European Union, will adopt US policies. This independence was evident in the conflicting NSABB and WHO assessments of H5N1 virus

---

[1] This is commonly referred to as the deficit model, where any lack of public support is assumed to be based on ignorance and the solution is more explanatory lectures (Mooney 2010). Meanwhile, surveys suggest educational attainment is not strongly correlated with support for controversial science policy questions (Funk, Rainie & Page 2015). While more widely accepted today, the first discussions of the underlying hubris of scientists (Feyerabend 1978) were treated as blasphemous. One could argue in our post-truth, 'fake news' modern world, the pendulum has swung too far the other way and science has lost its authoritative voice (Nichols 2017; Crease 2019). Yet the long-running General Social Survey finds Americans have had consistently high confidence in the scientific community for the past 40 years during a time when trust in most public institutions has steadily fallen. Climate change denial may be particularly alarming, but it is only one of many anti-science minority positions throughout history that has eventually collapsed under its own improbability. While the uncertain nature of knowledge is undeniable, so is the human tendency to overplay uncertainty when personally convenient.

research. Neither should the US expect the rest of the world to follow its lead given its own rather inconsistent commitment to multilateralism. For example, the US is the only United Nations member not party to the Cartagena Protocol on Biosafety or the more general Convention on Biological Diversity—two international agreements that regulate the transfer and use of living genetically engineered organisms.

Additionally, given the competitive nature of NIH funding and the decreasing cost of advanced labs, the proportion of life scientists in the US working independently of federal funding is growing. Newer philanthropic organizations, such as the Gates Foundation, are joining stalwarts like the Howard Hughes Medical Institute to provide a substantial source of research funding. Silicon Valley has also been luring life scientists from academia with the promise of better resources and an emphasis on results rather than grant-writing and publications.

The limited reach of US regulators within the global research community has been used as an argument for limited regulation—we cannot limit NIH-funded scientists for fear they will simply go elsewhere to perform their dangerous research. For example, China is expected to have multiple BSL-4 labs before 2025. This is a precarious argument. It represents the same Cold War attitude that seemed to compel an arms race of unquestioned nuclear weapons research. It also makes an implicit assumption of superior ethical sophistication the US science research community may no longer deserve (Sipp & Pei 2016).

While gain-of-function research on potentially pandemic pathogens is currently conducted solely in well-funded labs with strict biosafety measures, the issues that underlie this debate are even more serious in other research communities. For example, synthetic biology has a thriving do-it-yourself independent lab sub-culture (Keulartz & van den Belt 2016) of amateurs and entrepreneurs working without regulatory oversight, funding, or training in biological fields. The danger extends beyond a lack of minimal biosafety training to active high-risk attention-getting behavior (Baumgaertner 2018; Zhang 2018). In the absence of new regulations that restrict who can access this technology (or maybe even in spite of them), a tragedy and public backlash are nearly inevitable (Green 2016).

Engaging multiple perspectives

A common prescription for high uncertainty risk management is to encourage substantial public involvement. The public is critical to risk analysis because it serves as a source of 'moral imagination' (Johnson 1993), which lets us explore the consequences of our actions and imagine the plight of others (Coeckelbergh 2009). Conscious effort to enlarge the risk assessment process is necessary because of the pervasive myth of technocracy—the idea that only specialists are equipped to assess and manage technological risk decisions (Jasanoff 2016).

As an added benefit, a discursive approach that emphasizes mutual education can also be used for well-known risks the public consistently underestimates or overestimates (Renn & Klinke 2004).

However, the idealized model of public participation, which includes a well-informed and representative public, is difficult to achieve (Rossi 1997; Lavery 2018). Public engagement, while widely considered a prerequisite to good risk management, is not guaranteed to be useful or sufficient. Sometimes increasing the number of stakeholders in a decision process can lead to increased confusion and general intransigence. Citizen participation in technical policy decisions can be disappointing in practice because the participants start off overwhelmed by technical information and end up overwhelmed by their inability to judge conflicting expert assessments. Given the variety of epistemic and ethical value assumptions inherent to the risk assessment process, evaluating technological risk disputes is difficult enough for scientists already familiar with the technology, let alone the general public who must also absorb additional background information. One way to mitigate these issues is to provide science literacy mini-courses to public participants that discuss the types of scientific disputes and the contingent nature of science and scientific consensus (Reaven 1987).

Ironically, the need for scientific literacy training extends to scientists. While most working scientists have formal training in the methods and conduct of science, few are educated in its philosophical underpinnings. This is probably why some scientists are ambivalent or even dismissive when science policy issues arise. Despite the difficulties, it is important to engage the scientific community. To ensure accountable science, dissenting scientists must have open channels for voicing concerns where they need not fear retribution. Likewise, scientists should be encouraged to reflect on their work rather than assuming the existing regulatory process, such as institutional review boards, will catch any ethical or safety issues. Jennifer Doudna, one of the discoverers of CRISPR-Cas9, initially hesitated getting involved in the science policy surrounding her work: 'I told myself that bioethicists were better positioned to take the lead on such issues. Like everyone else, I wanted to get on with the science made possible by the technology' (Doudna 2015). Eventually, she changed her mind and advocated for scientists to become prepared for dealing with the larger implications of their work.

That said, the attention of experts tends to be concentrated on their area of expertise. Scientists will have a natural tendency to permissively promote science, while security experts will tend to conservatively restrict perceived risks (Klotz & Sylvester 2009; Selgelid 2009). Because the science research community is largely self-regulated (with notable exceptions in nuclear research), this tends toward minimal regulation. Although a robust dialogue within the scientific community is healthy, 'risk is more a political and cultural phenomenon than it is a technical one' and unelected 'privileged experts' should not dictate the terms of technological risk debates (Sarewitz 2015).

Conversely, self-regulation has some benefits. Security specialists are trained to focus on signs of malicious intent but may be less cognizant of unintended harm. Signs of impending accidents and other unintentional harms may best be detected by the members of the research community who are fully immersed in the field and understand the formal procedures and tacit knowledge associated with the research. This may be one reason that the NSABB, whose members are primarily scientists, became increasingly concerned with biosafety risk issues in the H5N1 debate even though the board was created to deal with biosecurity concerns (Fink 2004).

Despite a desire for a 'broad deliberative process,' the NSABB and NRC evaluations were largely headed by experts in virology and public health. Even the public comments were public only in the sense they came from outside of the organizations. For example, comments from the November 2014 NSABB meeting consisted of six public health specialists, a virologist, and a policy analyst. Furthermore, some of the most vocal experts had a vested personal interest in the outcome of the assessment. This led to some confusing science communication. For example, when first presenting their findings, Dr. Fouchier and others made bold claims that influenza transmissibility in ferrets and humans was nearly equivalent. However, after the public backlash and subsequent threats to the viability of the research, Dr. Fouchier made much weaker claims about the use of ferrets as human surrogates and downplayed the transmissibility and lethality of the engineered virus (Kahn 2012; Lipsitch & Inglesby 2015). Whether intentional or not, this appeared to be a case of miscalculation in risk communication (Sandman 2012). That is the initial announcement emphasized the danger of the H5N1 research to attract attention from peers and future funding. However, when the public heard the same message and panicked, subsequent media interactions attempted to minimize the danger of the work. This is a good example of the balancing act in dangerous science: scientists must appear to be cutting-edge while at the same time appearing safely incremental.

The concerns of narrow interests and bias led to calls for more active public engagement in the deliberative process; a reasonable request considering that most of the world's population was potentially affected by H5N1 research (Fineberg 2015; Schoch-Spana 2015). Of course, there is an asymmetry in global effects. The risks are the lowest and the potential benefits are the greatest to individuals in nations with the best access to modern health care (Quinn et al. 2011). Thus, the people with the most to lose and the least to gain were largely absent from the discussion. One letter to the NSABB chair voiced concerns the deliberative process did not include enough risk assessment experts, was not sufficiently international, did not have enough public comment opportunities, was generally opaque and moving too fast, and had an inherent conflict of interest by being funded by the NIH (Roberts & Relman 2015). While the NSABB acknowledged the need for more public input and many of the meetings were open to the public, the deliberations were poorly publicized and remained meetings of experts.

So how do we achieve public science?

Science policy making is a political and ethical process and should not be left entirely to the scientific community. 'Scientists may approach their research with the best of intentions and the good of society in mind, but it would be a mistake to assume that they could know what is best for the world—and a tragedy to foist that burden on them' (Evans 2013). So what is the solution? Engaging multiple perspectives increases the likelihood important considerations in risk assessment and viable options for risk management will not be missed. However, despite the global reach of modern science, truly inclusive public engagement at this scale is rarely attempted. The few previous attempts at international-scale public participation have failed to demonstrate any impact that justified the considerable effort, expense, and time involved (Rask 2013).

Some alternative forms of technology assessment have been proposed that do not rely solely on the scientific community. The most popular effort has been the creation of prediction markets (Mann 2016) that estimate the likelihood a technology will succeed or fail. Prediction markets could easily be adapted to estimate technological risk. A prediction market offers shares to the public that pay out only if an event occurs. The trading price of the shares reflect the collective belief of the likelihood of the event. As the trading price rises toward the payout amount, the estimated likelihood rises toward 100%. Likewise, a share price near zero reflects the public wisdom the event is unlikely to happen. Proponents of prediction markets appreciate its simple interpretation, the ease of participation, and its successes in predicting past political events compared to expert analysis or opinion polls.

However, critics of applying prediction markets to science policy have argued science rarely consists of distinct near-term events with clear outcomes like sports, an election, a daily stock price, or a quarterly earnings report. Thus, it is inherently difficult to sustain the large market needed to generate useful results; most investors are uninterested in tying up their money by placing bets on ambiguous events in the distant future. Likewise, prediction markets share the same manipulation issues as other financial markets. For example, a scientist privy to insider information could be tempted to game the market for personal financial gain. These shortcomings suggest prediction markets might actually make science policy worse (Thicke 2017). Another alternative, which uses predictions polls within forecasting tournaments (Tetlock, Mellers & Scoblic 2017), sidesteps the market manipulation issue, but the difficulty remains of trying to fit amorphous science policy into a method that works best for short-term, specific, and unambiguous events.[2]

---

[2] One could also argue short-term well-defined science is also the most likely to have commercial value. This is precisely the type of science that may already be getting outside scrutiny through venture capital or the stock market.

Another idea for broader technological risk assessment is to require scientists to obtain liability insurance for their research (Farquhar, Cotton-Barratt & Snyder-Beattie 2017). The strength of this approach is that it ignores the rather pointless exercise of attempting to quantify research benefits and concentrates on risks using a mature industry that specializes in risk assessment. Furthermore, requiring a financial outlay for potentially dangerous research would encourage responsible behavior—perhaps more than any other single action. However, the insurance industry does not have the best track record for estimating and adequately insuring a range of rare but substantial risks (for example, mortgage risk in 2008). Although many insurance companies offer policies for events that are difficult to estimate (terrorism, cyberattacks, etc.), the mere existence of these policies does not prove the risks are well-known or adequately insured. Ultimately, the largest issue is that scientific liability insurance rests on the premise there is a price for everything. But what if we cannot agree on a price due to conflicting interpretations of evidence or the research violates societal values that cannot be priced?

Outsourcing risk assessment to the financial industry brings up another important point common to all technological risk assessment. How do we engage the public so research does not further exacerbate inequality? It is well recognized the benefits of science and technology are heaped upon the residents of highly industrialized nations and trickle down to the rest of the world, if at all. Unfortunately, the risks of technological innovation are not distributed in the same fashion. The world's poorest are often disproportionately subjected to the highest environmental and health risks from modern science. So why do they get a small or non-existent seat at the table when technological risks are assessed?

The current model of public science assessment stems from the 1975 Asilomar conference on recombinant DNA. Less than 150 people—mostly scientists—convened at a small conference center on the Pacific coast to draw up safety guidelines for the new technology. The meeting focused on biosafety risk assessment, specifically excluding biosecurity issues or related social and ethical issues. Many scientists have pointed to Asilomar as a successful model for self-governance because of the subsequent lack of recombinant DNA catastrophes. Yet the scientific community regularly bemoans widespread public rejection of some of the results of this technology, such as genetically modified foods.[3] Proponents of the Asilomar model of science policy have missed the

---

[3] The public's 'anti-science' rejection of GMOs is not solely concerned about the techniques themselves, but rather specific modifications. To lump all GMOs together merely allows each side to selectively demonstrate the unreasonableness of the opposition. The same issue appears to be occurring with newer gene-editing techniques. This is a problem inherent to debating and regulating techniques rather than outcomes.

crucial point that there was a much larger discussion to be had and the public was the *most* important stakeholder. After all, what good is safely conducted science if you are not allowed to put it to use?

Subsequent controversial technologies have suffered the same fate. How many people even knew there was an International Summit on Human Gene Editing in Washington DC in December 2015? Probably a lot fewer than the number of people who cared about the topic. The second International Summit on Human Gene Editing in Hong Kong in November 2018 received a bit more media attention only because of the surprise announcement of the birth of gene-edited twins in China.

When the next controversy arises, will the scientific community once again feel victimized when protesters vilify their work when researchers are only trying to help humanity? To be fair, engaging the public can be exhausting. It takes up a lot of time that could be spent doing research. It can also be frustrating. For anyone even remotely familiar with the climate change debate in the US, it is clear some people choose to be willfully ignorant when their livelihood, lifestyle, political ideology, or religious beliefs are threatened. So public engagement does not promise success. However, skipping it practically guarantees failure.

That said, it is also important not to view public engagement as a bottomless hole of stupidity scientists must attempt to fill with superhuman patience. Yes, public opinions can often be painfully uninformed or seemingly illogical, but they can also be nuanced, insightful, and absolutely correct. It is also true there is a long history of scientists making claims in the past that turned out to be wrong, which have trained the public to be skeptical. In a democracy, open political debate remains our best method for separating the wheat from the chaff and arriving at a current best approximation of reality.

One proposed alternative to pseudo-public science policy is to create 'global observatories' (Jasanoff & Hurlbut 2018) for science controversies, which would serve as an information clearinghouse, track developments in the field, organize public meetings, and collect public survey data. The new model, inspired by the broad work of the International Panel on Climate Change, recognizes effective science policy is not a matter of just including a few academic bioethicists in a meeting of scientists. Rather, science must be widely and continuously discussed, like sports and weather. Compared to the ideal of the global observatory, the NSABB, with its narrow mission, may have been the wrong model for appropriate technical oversight of the avian influenza research debate.

A less technical alternative approach to expanding the sphere of consideration in risk management is through literature. The value of literature to improve empathy and moral imagination has long been argued and more recently supported by research (Nussbaum 1990; Bal & Veltkamp 2013; Kidd & Castano 2013). Philosopher Hans Jonas called for 'a science of hypothetical prediction' that acknowledged the impossibility of 'pre-inventing' technology, but realized that the reach of modern technology often exceeded our foresight.[4] He argued one of the best options was to cultivate an imaginative technological skepticism and suggested science fiction as a valuable source of inspiration for exploring the possible and laying the groundwork for risk management. More recently, the idea of 'science fiction prototyping' (Johnson 2011) has been proposed to create narrative scenarios of technological development that better engage the public. In the case of technological risk analysis, science fiction may suggest risk management options and provide fertile ground for moral imagination— a way to play out future technological scenarios and explore their ethical implications.

Inherent safety

Having discussed the various stakeholders and how to involve them in the science policy process, let us turn to what they should be discussing. A common reaction in modern risk management is to increase the resiliency of a system in the face of deep uncertainty (Linkov et al. 2014). Physical systems can be made more resilient by increasing the redundancy or independence of functions. But how do we make a line of research or technology more resilient when the potential risk to humanity is widespread or catastrophic? We could try to make society itself more resilient to the effects of research and emerging technologies, but it is not clear this is possible. It is also a rather unsettling and unrealistic approach—discussions of increasing human resiliency through population growth or by geophysical barriers, such as underground bunkers or space colonies, are not considered serious public policy (yet). For now, a better response to dangerous science is to reduce the risk. However, considering the apparent inevitability of technological progress, it would seem technological risk can only increase. What are we to do? Even if we accept the idea dangerous science will usually continue until it is discredited or something better comes along, we need not accept all research must be inevitable and unfettered. Between the extreme positions of outright banning technologies and fatalistic permissiveness lie the traditional methods of risk management, which generally fail to resolve research controversies. However, a more satisfying moderate risk management strategy exists within the concept of inherent safety.

Conventional risk management generally focuses on reducing the likelihood of an accident through safety procedures and equipment. In contract, the principle of inherent safety emphasizes the elimination of the hazard itself. The idea was first proposed in the chemical industry in a paper aptly titled, 'What you

---

[4] Jonas suggested one of the earliest forms of the precautionary principle: 'It is the rule, stated primitively, that the prophecy of doom is to be given greater heed than the prophecy of bliss' (Jonas 1984).

don't have, can't leak' (Kletz 1978). Inherent safety has been described as common sense, but not common knowledge. The idea is well-known in the chemical and nuclear engineering communities but not by scientists and engineers in other areas (Srinivasan & Natarajan 2012). In the nuclear engineering community, there is a different emphasis on the basic inherent safety principles due to the difficulty of reducing the hazardous material that is fundamental to the nuclear power process. Rather than reducing the material, the emphasis is on reducing hazardous systems (for example, designing a nuclear power plant that is incapable of a reactor core meltdown).

The distinction between reducing the hazard versus reducing the likelihood of the bad event has also been termed primary and secondary prevention, respectively (Hansson 2010a). By selecting safer materials during the early design phase, overall risk is reduced far more than by merely adding on safety systems or safety procedures as an afterthought. The basic principles of inherent safety consist of (1) minimizing the total amount of materials used, (2) substituting hazardous materials with alternatives that do not have the hazardous quality, (3) using materials under less hazardous conditions, and (4) simplifying processes to reduce errors (Kletz 1985; Khan & Amyotte 2003; Edwards 2005).

Another benefit of inherent safety is the ability to simultaneously address safety and security concerns. Terrorists are attracted to material hazards that can be exploited and safety systems that can be subverted. Inherently safe materials and systems discourage malevolent actors. Additionally, inherent safety is often simpler and less expensive than standard risk management efforts. Traditional approaches incur extra physical, procedural, and administrative layers that may include (Möller & Hansson 2008)

- safety reserves, where reserve capacity is included in the design (e.g., safety margins and safety factors);
- fail-safe design, where a system is designed to fail in a way that minimizes damage (e.g., safety barriers that isolate failure); and
- procedural safeguards (e.g., warning systems, safety audits, and safety training).

Historically, safety has been a secondary concern relegated to engineers at the production level instead of a primary concern at the early stages of research, when the most impact could be made (Edwards 2005). This is perhaps one reason the research community has not embraced the idea of inherent safety.

Unfortunately, the opportunities for inherent safety in research are only slightly more obvious in hindsight. In June 2014, laboratory workers were exposed to potentially viable anthrax bacteria at a CDC bioterrorism response lab. The subsequent internal investigation report found the exposure arose from testing whether a new mass spectrometer could quickly detect *B. anthracis*

(CDC 2014). However, the instrument manufacturer stated the instrument could not distinguish between virulent and relatively benign strains of the same species. Thus, using a low-risk non-pathogenic strain would have yielded the same results. Despite the obvious application of inherent safety, the CDC report recommendations still focused on traditional safety management, such as revised biosafety protocols and procedures; hazard reduction was mentioned only once and in the fifth of eight recommendations.

The findings of the report also reinforce two previous arguments. First, even at facilities with multiple safety barriers and extensive training, unanticipated errors do occur. This suggests traditional risk management solutions may offer false security by reducing the perception of risk more than the actual likelihood of bad events. Second, considering safety in the research design phase can often accomplish the same scientific goals while sidestepping controversy.

Ironically, a primary reason why inherent safety has been slow to catch on in industry is risk aversion (Edwards 2005). Most organizations are hesitant to fix things that do not appear to be broken—even systems with many near misses are considered to be safe until something too egregious to ignore occurs (Paté-Cornell & Cox 2014). Neither do organizations want to incur the potential risks of deviating from tradition and known practices. These same concerns are potential roadblocks to using inherent safety principles in dangerous science. By its nature, research is full of new ideas, but researchers often use well-established (and peer-reviewed) techniques. Thus, scientists may be just as hesitant as industry to incur the extra time and expense needed to consider inherent safety principles. Researchers who are focused on work efficiency may find the safeguards already in place to be onerous and may see any additional requests as superfluous. However, public pressure and threats to funding may lead scientists engaged in dangerous science to view inherent safety as a reasonable compromise. It is even possible inherently safe research could be faster if it allows cumbersome traditional safety systems to be avoided.

Inherent safety, like all human endeavors, is limited by knowledge and creativity. Invoking inherent safety first requires the realization a planned course of action is potentially unsafe. Likewise, inherently safe alternatives are not always obvious and may require considerable innovation. There is also the risk a design change may create a new unforeseen risk that did not previously exist. To avoid trading one potential risk for another unrecognized risk, it is useful to engage multiple perspectives to help offset limitations of imagination.

Given these caveats, it should be obvious safety is an iterative and never-ending process. The chemical industry, where the idea of inherent safety was first introduced, still struggles in the quest to minimize its negative impacts (Geiser 2015). Ultimately, the primary value of inherent safety is providing a complementary philosophical approach to standard probabilistic risk analytic thinking, which treats the hazard as a given (Johnson 2000).

Application to the H5N1 debate

While the biosecurity risks discussed in the H5N1 research debate were largely theoretical, the biosafety concerns were based on historical data and made a compelling case for the use of inherent safety principles. For example, the 1977 Russian influenza outbreak was an H1N1 strain genetically identical to a virus that caused a 1950 epidemic. The lack of mutations in the intervening 27 years suggested a frozen virus source—presumably from an unintended laboratory release (Webster et al. 1992; Lipsitch & Galvani 2014). Research accidents continue to occur even in the most modern and technologically advanced research facilities. A review of US reports of theft, loss, or release of select agents and toxins compiled by the CDC between 2004 and 2010 found no reports of theft, 88 reports of loss, and 639 cases of accidental release—14 of which occurred at BSL-4 facilities (Henkel, Miller & Weyant 2012).

While inherently safe research seems like commonsense, it is not emphasized in biosafety or biosecurity guides, which focus on formalized processes and training. The continued emphasis on traditional methods is unfortunate given the poor record of implementing such measures. For example, the European Committee for Standardization's CWA 15793 framework for lab biosafety was adopted in 2008, but by 2013 only one third of the European Biosafety Association's 118 members were using the framework and 15 percent were unaware of its existence (Butler 2014). Likewise, there has been insufficient effort to integrate biosecurity considerations into the culture of the life sciences. An informal survey of 220 graduate students and postdoctoral fellows at top NIH-funded US institutions found 80 percent of the respondents had bioethics or biosafety training, while only 10 percent had biosecurity training (Kahn 2012). Some theories for the dearth of biosecurity training include a lack of perceived relevance, a lack of familiarity, and a general reluctance to acknowledge life science research could be used to cause harm. Perhaps the largest issue is the general lack of awareness of biosecurity issues in the life science community (National Academies 2017a). It has been suggested the culture of safety in the life sciences could be most quickly improved by replicating the established models and hard-earned lessons of other fields (e.g., nuclear engineering) to create 'high-reliability organizations' (Trevan 2015; Perkins, Danskin & Rowe 2017).

Even after a string of incidents in the summer of 2014 placed renewed focus on biosafety, subsequent events suggested appropriate laboratory safeguards had yet to be implemented. In December 2014, it was discovered live Ebola virus was sent to the wrong lab within the CDC. In May 2015, it was discovered a US Army lab in Utah had likely shipped live, rather than inactivated, B. anthracis samples to approximately 30 US and foreign labs over several years. This suggested reducing human error was difficult even when biosafety issues were in the headlines.

Early in the debate, critics cited the 1947 Nuremberg Code, a foundational bioethics document, to argue most gain-of-function research on potentially pandemic pathogens was not ethically defensible because the same benefits could be attained by safer alternatives (Lipsitch & Galvani 2014). Specifically, the two main goals of these experiments, guiding vaccine development and interpretation of surveillance data, can be achieved by other methods, such as molecular dynamical modeling, *in vitro* experiments that involve single proteins or inactive viral components, and genetic sequence comparison studies (cf. Russell et al. 2014). Likewise, the goal of reducing influenza pandemics could be more safely pursued by research on universal flu vaccines (Impagliazzo et al. 2015), broad-spectrum antiviral drugs (Shen et al. 2017), and improved vaccine manufacturing. All these alternatives fall within the general concept of more inherently safe research, although there is debate over the viability of some avenues, such as universal flu vaccine research (see Cohen 2018; Nabel & Shiver 2019). Ultimately, the narrow set of questions that can *only* be answered by gain-of-function research may have limited public health benefits (Lipsitch 2018).

While the initial HHS guidelines for regulating gain-of-function research did include the concept of inherent safety in its third criterion (Patterson et al. 2013), the discussion among scientists and regulators was focused on traditional risk management. However, suggestions to incorporate inherent safety slowly gained support as the debate progressed. Eventually, even Dr. Kawaoka, the lead scientist for one of the original H5N1 papers, conceded some research could use alternative techniques, such as loss-of-function studies, less pathogenic viruses, or analyses of observable traits (NRC 2015). This is an admirable evolution of perspective considering researchers spend substantial effort to become experts in particular techniques; one should not expect them to instantly abandon their prior work when potentially better options are presented.

Because the application of inherent safety is still relatively new to the biosciences, there are some special considerations. First, the principles of inherent safety are more restrictive than when applied to the fields of chemical and nuclear engineering because of the unique ability of biologically hazardous materials to replicate. Thus, inherent safety may require a hazardous organism be eliminated, rendered unable to reproduce, or have survival limitations imposed. This last approach has been pursued in synthetic biology through genetically modified organisms that only survive in the presence of an anthropogenic metabolite, such as a synthetic amino acid (Mandell et al. 2015).

That said, many applications of inherent safety in the biosciences do not require such extreme innovation. For example, certain types of attenuated viruses were found to have the potential to recombine into more virulent forms when the same livestock population was vaccinated with two different forms of the same vaccine (Lee et al. 2012). Traditional biosafety risk management would suggest stricter policies and processes are needed to prevent double vaccinations. Meanwhile, an inherent safety approach would attempt to reformulate the vaccine using an inactivated rather than an attenuated virus to remove the potential hazard altogether.[5]

Improving Risk-Benefit Assessments

The second chapter of this book compares several approaches to valuing the benefits of research to show how they can yield disparate assessments. Likewise, the third chapter details some of the many value assumptions inherent in risk assessments that prevent claims of objectivity. These lines of argument help explain why risk-benefit analysis is more appropriate for evaluating well-defined problems with copious data but has limited utility for resolving debates over dangerous science where uncertainty is high and data are sparse. Does this mean risk-benefit analysis has no place in assessing and managing dangerous science? No, it means a risk-benefit assessment, as commonly used to assess dangerous research, is ineffective as a primary decision criterion. However, we can considerably improve its utility by following a few recommendations.

Recommendation 1: Change expectations

While the purpose of risk assessment is understood in theory, it is less so in practice (Apostolakis 2004). The effort of generating a time-consuming and costly risk-benefit assessment often carries an expectation the results will translate into a decision. However, expecting all risk-benefit assessments to yield clear-cut, consensus-building answers is unrealistic. The controversies surrounding some lines of research are based on fundamental disagreements over ethics and appropriate public risk that will not change due to the findings of a formal risk assessment. Certain issues are more amenable to analysis than others for reasons already noted. Uncontroversial and well-defined questions can essentially use an assessment as a decision tool, but more controversial and complex questions should treat a risk-benefit assessment as a risk exploration tool. While this may seem unhelpfully vague, it simply means once an assessment itself becomes a source of controversy, more data will probably not resolve disagreement. Controversy is a sign to switch expectations from decision criteria to risk exploration tool. In fact, a thorough risk assessment may actually increase controversy because it uncovers previously overlooked risks.

---

[5] Vaccine development has also been suggested as a more inherently safe method of combating antimicrobial resistance in common pathogens. Antibiotics are difficult to develop and lose effectiveness over time as resistance spreads, while vaccines are easier to develop and vaccine resistance is rare (Rappuoli, Bloom & Black 2017).

Furthermore, any request to conduct a risk assessment will not be accompanied by a well-defined procedure because there is no consensus on what constitutes the proper accounting of benefits or what underlying value assumptions should be used to estimate risk. By concentrating on the correctness of an assessment, stakeholders merely end up creating a secondary debate. The simple solution is to make sure that all stakeholders—policymakers, analysts, scientists, and the public—understand formal risk-benefit assessments inform the decision process but are not the decision process itself. A quality risk-benefit assessment should provide insight within the broader scientific and political debate.

The H5N1 research debate serves as a classic example of when to forego the traditional use of risk assessment as a decision tool. Before the analysis was conducted, expert opinion on the annual likelihood of a research-induced pandemic ranged from better than 1 in 100 to less than 1 in 10 billion. This wild disparity among opposing experts guaranteed the outcome of any single formal risk assessment was also going to be contentious. Likewise, there was no consensus in the scientific community regarding the basic value of the research. Some scientists claimed the work was critical to public health and relatively low-risk (Morens, Subbarao & Taubenberger 2012; Palese & Wang 2012), while others claimed the approach had no predictive ability useful to pandemic preparedness and presented an unnecessary public health risk (Mahmoud 2013; Rey, Schwartz & Wain-Hobson 2013). Wisely, some individuals raised doubts early on that a risk-benefit assessment would resolve the debate (Uhlenhaut, Burger & Schaade 2013; Casadevall & Imperiale 2014). The NSABB, which had formerly noted risk-benefit assessments were subjective (NSABB 2007), originally discussed the H5N1 assessment in the terms *subjective* and *objective* but later reframed its expectations by switching to *qualitative* and *quantitative*. Anyone who went into the process believing a risk-benefit assessment was a consensus-building tool was bound for disappointment.

Recommendation 2: Use broad uncertainty assumptions

Uncertainty is inherent to all risk-benefit assessments, and its characterization is critical to an assessment's utility. Even for risk-benefit assessments that address fairly narrow questions, it is best to use a broad conception of uncertainty. This includes acknowledging not all uncertainty can be represented probabilistically without making important assumptions, dependencies among parameters are not always known, and the correctness or completeness of the model is often ambiguous. The value assumptions roadmap outlined in the third chapter can be used as a guide. Even in a highly quantitative assessment, broader notions of uncertainty can still generate informative results; more importantly, the results will not claim certainty that does not exist.

Recommendation 3: Use multiple methods

When resources allow it, assessments will benefit from using multiple techniques developed from risk analysis. Some examples would be to use both a Monte Carlo simulation and probability bounds analysis together or to have a narrow quantitative assessment accompanied by a broad qualitative assessment. The work in the second and third chapters provide guidance on the range of methods available. The value of multiple methods is to give the proponents of various metrics and statistical techniques (based on their epistemic preferences) a chance to feel confident about the results of the assessment. Likewise, individuals without any methodological preference will be comforted by the thoroughness of the analysis and the range of analytic frameworks.

Recommendation 4: Use the analysis to design better research

If analysts follow the first three recommendations, a risk-benefit assessment will be better suited for exploring risk. This can help researchers rethink their work to accomplish their research goals by inherently safer or more ethical means. Ideally, the best opportunity to resolve a debate over dangerous science is to apply the inherent safety design concept to avoid the debate altogether.

As previously mentioned, this is no easy task. It requires a new mindset and a cultural shift within the communities of scientists and engineers. If we are to embrace the idea of inherent safety, the conception of every research project must start with a foundation of humility onto which we build our clever ideas. Whenever we ask, 'Can this be done?' we must also ask, 'Will it fail safely when something goes wrong?'

Science policy done better: Gene drives

Given the many examples in which poor science policy seems to be failing us, it is useful to look at a positive example of science risk assessment and management. One recent example is the work on gene drives. The basic concept of a gene drive is a genetic mutation that is nearly universally inherited and can spread quickly through a species. While they are known to naturally occur in some species, bioengineers now have the ability to create gene drives that accomplish specific tasks. While a gene drive could be used to accelerate a genetic study in the lab, a bolder and more controversial application of gene drives is to eliminate vector-borne diseases by mutation or eradication of their host species. Given the potential danger of such efforts, some scientists involved have been proactive about their work and have sought formal risk assessment and public input before proceeding too far. This has been a challenge considering the speed with which the science is developing. The first suggestion of a CRISPR-Cas9-mediated gene drive for species alteration was published in 2014 (Esvelt et al. 2014). A paper demonstrating an (inadvertent) CRISPR gene drive in laboratory fruit flies was published the following year (Gantz & Bier 2015), and a National Academies report on gene drive policy was published in 2016 (National Academies 2016).

Gene drives appear to be a particularly powerful biotechnology with endless ecological applications. A few proposals include making white-footed mice, the primary vector for Lyme disease, immune to the bacterium; eradicating all invasive rats, possums, and stoats from New Zealand; and eradicating invasive spotted knapweed and pigweed from the US. However, the most frequently discussed use is the mutation or elimination of some species of mosquito.

The benefit to humans of eliminating mosquitoes is obvious. They are the primary vector for a multitude of diseases, including malaria, yellow fever, dengue fever, West Nile virus, chikungunya, Zika, and several types of viral encephalitis. According to WHO estimates, malaria alone results in approximately half a million deaths (mostly children) each year. However, any proposal to drive a species to extinction is inherently controversial. While many biologists believe the ecological niche occupied by mosquitoes would be quickly filled by other insects (Fang 2010), mosquitoes provide an ecological service as pollinators and are an important food source for many species (Poulin, Lefebvre & Paz 2010). The substantial biomass mosquitoes comprise in some arctic and aquatic environments hints at the potential impact of their absence. Likewise, the concern a gene drive mechanism might be transferred to another species and drive an unintended target to extinction is reason enough to proceed with great caution and has already prompted formal risk-benefit assessments (e.g., Hayes et al. 2015).

If we view such reports with the ultimate purpose of spurring us to innovate more inherently safe research and technologies, we can see the value of gene drive risk assessments. Initial discussions centered on using a gene drive engineered to destroy X chromosomes during sperm development (Galizi et al. 2014). This results in a male-dominant trait that is passed on to subsequent generations and eventually drives the species to extinction when there are no more females to reproduce. Since then, the self-perpetuating extinction drive has become even more efficient with extinction likely within a dozen generations (Kyrou et al. 2018). However, other options that might have fewer unintended consequences have also emerged. Some of these less drastic suggestions include genetically engineering mosquitoes to be immune to the malaria parasite (Gantz et al. 2015); using two strains of the common bacterial parasite *Wolbachia* to infect male and female mosquitoes, which prevents successful reproduction; creating male mosquitoes that carry a gene that prevents offspring from maturing (Phuc et al. 2007; Windbichler et al. 2007); or using *Wolbachia* to prevent disease transmission by female mosquitoes. The last three alternatives have already been tested in the real world with promising results but have very different risk-benefit profiles. The genetic modification solution is self-limiting, reduces the overall mosquito population, and requires expensive redeployment. Meanwhile, the second *Wolbachia* solution is self-perpetuating, maintains the local mosquito population, and presently exists in nature (Servick 2016). The public is already starting to see these alternatives as preferable to the status quo of using insecticides, which have considerable ecological impacts and efficacy issues as resistance develops.

Conversely, there is a growing concern among gene drive researchers that the technology is not powerful enough. The CRISPR-Cas9 system is known to occasionally make random DNA additions or subtractions after a cut. If the cut is sown back together before substitution, then it is no longer recognized as a gene splicing site. Likewise, natural genetic variation within species means some individuals within the target species will be immune to the gene drive substitution (Hammond et al. 2016). Over time, resistance to a gene drive should be naturally selected. Discovery of a variety of anti-CRISPR proteins (Acrs) widely found in bacteria and archaea also suggests there are natural defenses to the technology. Early work using mouse embryos has suggested constructing a consistent gene drive in mammals is more difficult than in insects but nonetheless achievable (Grunwald et al. 2019).

However, biologists Keven Esvelt, an originator of the CRISPR-mediated gene drive concept, argues mathematical modeling of first-generation gene drives demonstrates they are highly invasive in nature and need to be modified before even field trials should proceed (Esvelt & Gemmell 2017). Compounding the complexity of the gene drive discussion is the wide range of potential uses. An aggressive gene drive may be desirable if the intent is to rid a remote island of a destructive invasive species but is a serious cause for concern if introduced, for example, in a marine environment where there is high potential for the drive to spread globally.

The work of Dr. Esvelt and colleagues is particularly praiseworthy in its foresight to proactively address the potential dangers of gene drives for unintended consequences or even for weaponization by malevolent actors. Some of these precautions include creating secondary anti-gene drives that can undo other drives (albeit imperfectly), developing gene drives split into two parts that are not always inherited together to slow the process, or using gene drives separated into even more distinct components that operate in a sequential daisy-chain but eventually burn out as stages of the chain are lost through successive generations.

These technical safeguards are accompanied by perhaps an even more important contribution to science policy: pushing the scientific community toward open discussions among researchers and the public. Much of Dr. Esvelt's work involves sincere dialogue with potential test communities (Yong 2017) and fellow scientists in what amounts to a pragmatic version of 'real-time technology assessment' (Guston & Sarewitz 2002). Dr. Esvelt's efforts are backed by the National Academies report on gene drives, 'Experts acting alone will not be able to identify or weigh the true costs and benefits of gene drives' (National Academies 2016). This is a difficult undertaking given the existing competitive structure of science funding, which unintentionally encourages keeping secrets until publication is ready. While the gene drive research community remains small, cooperation is possible. It remains to be seen if the growing field will eventually revert to standard scientific secrecy and the dangers that go with it. Furthermore, a primary source of funding for early-stage gene drive research has been the US military (Defense Advanced Research Projects Agency), where open public review does not come naturally.

The ultimate success of Dr. Esvelt's efforts to reform science remain to be seen. However, regardless of whether one considers him to be on the vanguard of a new era of open public science or just the latest in a long line of quixotic warriors tilting against the unyielding edifice of science culture, it is important to understand he is absolutely correct. Hopefully, this book provides the convincing detailed arguments for why scientists and engineers should follow his lead.

Final Thoughts on Dangerous Science

A primary theme of this book is no detailed risk-benefit methodology to assess dangerous science exists because no consensus can be found regarding exactly how to categorize and quantify the risks and benefits of any controversial science program. The various value judgments inherent in the process tend to yield an assessment that is controversial itself. The practical advice here is to simply temper expectations of risk-benefit assessment as a quantitative decision tool: it is better viewed as a tool for exploring and communicating the social impact of research and technology. The discussion of the practical and philosophical difficulties underlying risk-benefit assessment in this book should help scientists and engineers perform assessments that better clarify specific areas of consensus and disagreement among researchers, the public, and policymakers. Likewise, formal assessments should not be used as a stalling tactic to distract concerned citizens.

Another theme is that in the absence of a convincing assessment, pre-existing technological risk attitudes guide management decisions. I note a general trend of permissive management of dangerous science and suggest, when possible, redesigning research to avoid controversy is the most viable way forward. Lessons learned from the case study of controversial H5N1 avian influenza virus research can be quite useful when presented to a larger audience. The research community is not monolithic; insight from past events can be quickly forgotten, and the best practices in one field are often ignored by other fields that face similar problems but have little interaction. The following are a few additional lessons we can glean from the H5N1 debate.

Post hoc *management is difficult*

Once funding has been obtained and work has begun, researchers have more than a purely intellectual position regarding a line of research. Funding creates an immediate personal financial interest as well as a long-term career impact. As Upton Sinclair said, 'It is difficult to get a man to understand something when his job depends on not understanding it.' Once work has begun, the investment of time and effort creates an emotional attachment. Researchers will defend their work for reasons that may have more to do with personal reputation and feelings of ownership than with the merits of the work itself. Expecting researchers to self-govern their work is asking a lot of even the most well-intentioned scientists.

Furthermore, *post hoc* management options are limited. Both openness and secrecy approaches to science dissemination can have unintended consequences (Lewis et al. 2019). For potentially dangerous research, censorship is commonly proposed. However, the ubiquity of modern communications technology makes effective censoring increasingly difficult. Moreover, past censorship, mostly in the form of classification for national security purposes, has an unpleasant history of shielding scientists from essential public oversight (Evans 2013). Openness can reduce ethical issues when researchers become less inclined to be associated with research that is publicly criticized. It is also unclear whether restricting communication is a useful technique for reducing technological risk. Open debate engages multiple perspectives and improves the likelihood potential accidents or unintended consequences will be discovered.

Conversely, once research is published, other scientists may attempt to replicate the work or use the results to perform similar research, thereby increasing the overall risk. If the work is potentially dangerous, publication may be encouraging other laboratories to also perform dangerous work. It would seem advances in science, yielding increasingly powerful technology, challenge our ability to maintain a free society. However, many of these issues do not arise when considering proactive regulation at the funding stage.[6]

---

[6] Of course, *post hoc* management is sometimes unavoidable. The discovery of the first new *Clostridium botulinum* toxin in 40 years prompted a redacted 2013 paper in the *Journal of Infectious Diseases* due to dual-use concerns. This was a major event because the botulism-causing neurotoxic proteins are the most lethal substances known—fatal doses are measured in nanograms. Later it was discovered the new toxin responded to existing antitoxins and the full details were published. Proponents of the US policy on dual use research of concern believed this was an example of the policy working. Critics argued public health was risked by slowing down the dissemination of important information to other researchers who could have discovered the existing treatment faster.

Research controversies are predictably unpredictable

Research controversies are like snowflakes: unique in detail, but identical from a distance. The principle of dangerous science outlines a general trend in research controversies. With few exceptions, a line of ethically questionable or potentially dangerous research will continue until it has been found to be unproductive or inferior to an alternative. However, while there are general trends in the evolution of research controversies, the outcomes are contingent upon the specific technical details, utility, and viable alternatives. In this regard, the principle of dangerous science has limited predictive power. However, predictive power is not the only valuable characteristic of a theory.[7] Rather, the principle is a simple framework for interpreting a situation that frequently arises in science policy and technology assessment.

Dangerous Science abounds

The avian flu gain-of-function research debate is a prime example of low-embodiment, low-governability science and technology. However, other examples are plentiful. For instance, cybersecurity research has even lower material requirements (a cheap computer and internet access) and represents a potential threat that is almost exclusively information-driven. In an increasingly integrated world, critical infrastructure (e.g., utilities and transportation systems) are at risk of disruption. The potential economic and national security risks only increase as more systems become automated and interconnected. It is no surprise the field of cybersecurity is experiencing explosive growth. Cyberwarfare is also full of unique ethical situations. For example, the malicious code Stuxnet, which is believed to have destroyed approximately 1000 Iranian uranium enrichment centrifuges, has been argued to be the first highly targeted ethical weapon of war (Singer & Friedman 2014).

This in no way minimizes the continued importance of traditional material-centric dual-use policy. The old threats have not disappeared, but they must now vie for our attention among a growing crowd of technological risks. This discussion of the role of risk-benefit analysis also applies to technologies with high material embodiment. For example, a relatively new uranium enrichment technique using lasers was lauded by the commercial nuclear power industry but created considerable alarm in the nuclear non-proliferation community because it would make uranium enrichment much easier. When the US Nuclear Regulatory Commission (NRC) licensed this technology in 2012, opponents

---

[7] For example, game theory, originally lauded for describing economic systems, has lost some of its luster in recent years due to its lack of predictive power (Rubinstein 2006). Nonetheless, it still remains a valuable framework for analyzing certain problems and has generated useful niche applications, such as evolutionary game theory.

criticized the commission's narrow conception of risks—considering only the physical security of an enrichment facility while ignoring the broader societal implications of encouraging the technology.[8]

Likewise, entirely new fields of controversial material-centric research are emerging. For example, the current debate over using geoengineering to mitigate climate change has raised concerns regarding unintended climate consequences and the moral hazard of using geoengineering as an excuse to delay implementing more permanent and difficult mitigation strategies.

Where do we go from here?

The assessment and management of dangerous science is situated within a larger group of questions. When controversial research or technology becomes publicly known, questions are invariably asked regarding who funded the work, why it was allowed, and so on. Considering the examples of controversial research programs previously discussed, an important question arises: in a free and open society, how, if at all, should we control science research and technology development? To adequately address this question, we are led to a series of secondary issues partially addressed here: who should make the decisions (as many stakeholders as practicable), what form should oversight take (eliminating the hazard is preferred to using safety equipment and procedures), and at what stage should controls be established (as early as possible)? However, the implementation of ideal science policy remains elusive.

The most dangerous emerging and existing technologies are the ones we do not question. The long history of synthetic chemicals and products no longer commercially available due to recognized public health risks attests to the value of proactively assessing risk[9]—science research is no different. When scientists rush headlong into their research, often propelled by the competitive urge for first publication, we may be unpleasantly surprised on occasion. Our saving grace is most scientists spend a great deal of time considering their research and generally try to be deliberate and thoughtful in their work. However, despite the obvious need for scientists to consider the social and ethical impacts of their research, they are often actively discouraged from doing so. For example, when research on the *de novo* synthesis of the poliovirus was published in 2002, the editors of *Science* insisted on removing any discussion of ethical and social implications from the final article (Wimmer 2006). Without any reassurance

---

[8] The NRC's potential bias toward industry-friendly limited risk analysis has been noted elsewhere (Jaczko 2019).

[9] The widespread use of microbeads is a good example. It took many years to recognize the negative health and environmental impacts of micro-plastics and then to ban their use in personal care products. Some basic forethought could have prevented a lot of unnecessary harm.

that thoughtful humans were conducting the research, it is no surprise the public assumed the worst.

That said, self-reflection is not enough. It is unreasonable to ask scientists and engineers to advance their field and their careers while also dispassionately evaluating the potential social impacts of their work. They know a lot about their research and thus often have the most informed and nuanced opinions on its implications. However, their knowledge is deep but narrow and biased by personal interests. It may be better than most, but it is not enough.

The old fantasy of detached scientists working in ivory towers falls apart in light of modern technology. It is increasingly possible that a few highly trained and well-funded scientists could, for example, regionally wipe out an entire species. While this might be a reasonable idea in specific cases, the gravity of such a solution requires a thorough exploration of the risks and implications. Just as you cannot un-ring a bell, some science is irreversible (although the de-extinction movement would argue otherwise). We should be sure of not only our intentions, but also the consequences of our actions.

If unquestioned science is the most dangerous science, then open discussion involving many stakeholders is the solution. Many opinions from diverse backgrounds generally improves risk assessment. This is one of the best reasons for public funding of research and development. When the public funds research, more assessment and oversight is likely at the early stages when thoughtful discussion can be most productive. This is not to say privately funded research is necessarily dangerous, but public funding at least gives us a better chance of having a say in what direction science is heading. Given the power of science and technology in modern society, engaged public oversight is an essential requirement of any truly functional democracy.

Afterword

Substantial portions of this book's critique of formal risk-benefit assessment have been rather academic and theoretical, so let us end with some humbling real-world examples of why we need robust public conversations about science and technology policy. The following are ten informal assessments of various technologies made in the last century. All the assessments are made by accomplished scientists, engineers, or inventors speaking in their field of expertise or about their own work. This list is not intended to single out any individuals for ridicule—imperfect foresight is a universal affliction. Rather, the point here is to remind ourselves individual experts often lack the emotional distance, range of experience, or even the time to fully imagine and consider the consequences of their creations.

'When my brother and I built and flew the first man-carrying flying machine, we thought we were introducing into the world an invention which would make further wars practically impossible. That we were not alone in this thought is evidenced by the fact that the French Peace Society presented us with medals on account of our invention. We thought governments would realize the impossibility of winning by surprise attacks, and that no country would enter into war with another of equal size when it knew that it would have to win by simply wearing out the enemy.'

—Orville Wright, June 1917 letter discussing the use of airplanes in WWI

'So I repeat that while theoretically and technically television may be feasible, yet commercially and financially, I consider it an impossibility; a development of which we need not waste little time in dreaming.'
—Lee de Forest, inventor and pioneer of
radio and film technologies, 1926

'There is not the slightest indication that [nuclear energy] will ever be obtainable. It would mean that the atom would have to be shattered at will.'
– Albert Einstein, quoted in the *Pittsburgh Post-Gazette* in 1934

'There is practically no chance communications space satellites will be used to provide better telephone, telegraph, television or radio service inside the United States.'
– T.A.M. Craven, Navy engineer, radio officer and
Federal Communications Commission (FCC)
commissioner, in 1961

'Cellular phones will absolutely not replace local wire systems. Even if you project it beyond our lifetimes, it won't be cheap enough.'
– Marty Cooper, Motorola engineer and developer of the first cell
phone, in 1981 *Christian Science Monitor* interview

'I predict the Internet will soon go spectacularly supernova and in 1996 catastrophically collapse.'
– Robert Metcalfe, co-inventor of Ethernet and
founder of 3Com, writing in 1995

'The subscription model of buying music is bankrupt. I think you could make available the Second Coming in a subscription model, and it might not be successful.'
– Steve Jobs, co-founder of Apple, in a 2003
Rolling Stone interview

'Spam will soon be a thing of the past.'
– Bill Gates, co-founder of Microsoft, claimed spam would be
solved in two years according to a January, 2004
BBC interview at the World Economic Forum

'The probability of having an accident is 50 percent lower if you have Auto-pilot on. Even with our first version, it's almost twice as good as a person.'
– Elon Musk, CEO of Tesla, in April 2016 referring to Tesla's autonomous driving software only months before an Autopilot-driven Tesla drove into the side of a turning cargo truck, resulting in the world's first self-driving vehicle fatality.

'I think the idea that fake news on Facebook, of which is a very small amount of the content, influenced the election in any way is a pretty crazy idea.'
– Mark Zuckerberg, CEO of Facebook, in a November 2016 Techonomy conference interview referring to the 2016 US presidential elections

References

Adams, J. D., & Sveikauskas, L. (1993). *Academic Science, Industrial R&D, and the Growth of Inputs.* Washington, DC: Bureau of the Census.

Adleman, L. (2014, October 16). Resurrecting Smallpox? Easier Than You Think. *New York Times.*

Ahmadpoor, M., & Jones, B. F. (2017). The dual frontier: Patented inventions and prior scientific advance. *Science, 357*(6351), 583–587. DOI: https://doi.org/10.1126/science.aam9527

Alberts, B. et al. (2014). Rescuing US biomedical research from its systemic flaws. *Proceedings of the National Academy of Sciences, 111*(16), 5773–5777. DOI: https://doi.org/10.1073/pnas.1404402111

Alikani, M. et al. (2017). First birth following spindle transfer for mitochondrial replacement therapy: Hope and trepidation. *Reproductive BioMedicine Online, 34*(4), 333–336. DOI: https://doi.org/10.1016/j.rbmo.2017.02.004

Althaus, C. E. (2005). A disciplinary perspective on the epistemological status of risk. *Risk analysis, 25*(3), 567–88. DOI: https://doi.org/10.1111/j.1539-6924.2005.00625.x

Anjum, R. L., & Rocca, E. (2019). From Ideal to Real Risk: Philosophy of Causation Meets Risk Analysis. *Risk Analysis, 39*(3), 729–740. DOI: https://doi.org/10.1111/risa.13187

Apostolakis, G. E. (2004). How useful is quantitative risk assessment? *Risk analysis, 24*(3), 515–20. DOI: https://doi.org/10.1111/j.0272-4332.2004.00455.x

Arrow, K. et al. (2013). Determining Benefits and Costs for Future Generations. *Science, 341*(6144), 349–350. DOI: https://doi.org/10.1126/science.1235665

Arthur, W. B. (2009). *The Nature of Technology: What it is and How it Evolves.* New York: Simon & Schuster/Free Press.

Assaad, F. A. et al. (1980). A revision of the system of nomenclature for influenza viruses: A WHO Memorandum. *Bulletin of the World Health Organization, 58*(4), 585–591.

van Asselt, M. B. A., & **Rotmans, J.** (2002). Uncertainty in Integrated Assessment Modelling. *Climatic Change, 54*(1–2), 75–105. DOI: https://doi.org/10.1023/A:1015783803445

Attenberg, J., Ipeirotis, P., & **Provost, F.** (2015). Beat the Machine: Challenging Humans to Find a Predictive Model's 'Unknown Unknowns'. *Journal of Data and Information Quality (JDIQ), 6*(1), 1. DOI: https://doi.org/10.1145/2700832

Aven, T. (2008). *Risk Analysis: Assessing Uncertainties Beyond Expected Values and Probabilities.* New Jersey: John Wiley & Sons.

Aven, T. (2010). On how to define, understand and describe risk. *Reliability Engineering & System Safety, 95*(6), 623–631. DOI: https://doi.org/10.1016/j.ress.2010.01.011

Aven, T. (2015). On the allegations that small risks are treated out of proportion to their importance. *Reliability Engineering & System Safety, 140,* 116–121. DOI: https://doi.org/10.1016/j.ress.2015.04.001

Aven, T., & **Krohn, B. S.** (2014). A new perspective on how to understand, assess and manage risk and the unforeseen. *Reliability Engineering and System Safety, 121,* 1–10. DOI: https://doi.org/10.1016/j.ress.2013.07.005

Aven, T., & **Renn, O.** (2009). On risk defined as an event where the outcome is uncertain. *Journal of Risk Research, 12*(1), 1–11.

Aven, T., & **Renn, O.** (2010). *Risk Management and Governance: Concepts, Guidelines and Applications.* Berlin, Germany: Springer Science & Business Media.

Aven, T., & **Zio, E.** (2014). Foundational issues in risk assessment and risk management. *Risk analysis, 34*(7), 1164–72. DOI: https://doi.org/10.1111/risa.12132

Azoulay, P., Fons-Rosen, C., & **Zivin, J. S. G.** (2015). *Does Science Advance One Funeral at a Time?* Working Paper 21788. National Bureau of Economic Research. DOI: https://doi.org/10.3386/w21788

Badash, L. (1972). The Completeness of Nineteenth-Century Science. *Isis, 63*(1), 48–58. DOI: https://doi.org/10.1086/350840

Bal, P. M., & **Veltkamp, M.** (2013). How does fiction reading influence empathy? An experimental investigation on the role of emotional transportation. *PloS ONE, 8*(1), e55341–e55341. DOI: https://doi.org/10.1371/journal.pone.0055341

Balachandra, R., & **Friar, J. H.** (1997). Factors for success in R&D projects and new product innovation: A contextual framework. *IEEE Transactions on Engineering Management, 44*(3), 276–287.

Balch, M. S. (2012). Mathematical foundations for a theory of confidence structures. *International Journal of Approximate Reasoning, 53*(7), 1003–1019. DOI: https://doi.org/10.1016/j.ijar.2012.05.006

Baltimore, B. D. et al. (2015). A prudent path forward for genomic engineering and germline gene modification. *Science*. DOI: https://doi.org/10.1126/science.aab1028

Ban, T. A. (2006). The role of serendipity in drug discovery. *Dialogues in Clinical Neuroscience, 8*(3), 335–344.

Baum, A., Fleming, R., & Davidson, L. M. (1983). Natural Disaster and Technological Catastrophe. *Environment and Behavior, 15*(3), 333–354. DOI: https://doi.org/10.1177/0013916583153004

Baumgaertner, E. (2018, May 14). As D.I.Y. Gene Editing Gains Popularity, 'Someone Is Going to Get Hurt.' *The New York Times*. Retrieved from https://www.nytimes.com/2018/05/14/science/biohackers-gene-editing-virus.html

van den Belt, H. (2009). Playing God in Frankenstein's Footsteps: Synthetic Biology and the Meaning of Life. *NanoEthics, 3*(3), 257. DOI: https://doi.org/10.1007/s11569-009-0079-6

Ben-Haim, Y. (2006). *Info-Gap Decision Theory: Decisions Under Severe Uncertainty*. 2nd ed. Oxford: Academic Press.

Ben-Haim, Y. (2012). Doing our best: Optimization and the management of risk. *Risk analysis, 32*(8), 1326–32. DOI: https://doi.org/10.1111/j.1539-6924.2012.01818.x

Berg, P., & Singer, M. F. (1995). The recombinant DNA controversy: Twenty years later. *Proceedings of the National Academy of Sciences, 92*(20), 9011–9013. DOI: https://doi.org/10.1073/pnas.92.20.9011

Berger, E. (2013, May 17). Is NASA about jobs, or actually accomplishing something? *Houston Chronicle*.

Bernal, J. D. (1939). *The Social Function of Science*. London: George Routledge & Sons Ltd.

Bessen, J. (2008). The value of U.S. patents by owner and patent characteristics. *Research Policy, 37*(5), 932–945.

Bimber, B. (1994). Three faces of technological determinism. In M. R. Smith, & L. Marx (Eds.), *Does Technology Drive History: The Dilemma of Technological Determinism* (80–100). Cambridge, MA: MIT Press.

Blackwell, D. (1951). Comparison of Experiments. *Second Berkeley Symposium on Mathematical Statistics and Probability, 1*, 93–102.

Boddie, C. et al. (2015). Assessing the bioweapons threat. *Science, 349*(6250), 792–793. DOI: https://doi.org/10.1126/science.aab0713

Boholm, M., Möller, N., & Hansson, S. O. (2016). The Concepts of Risk, Safety, and Security: Applications in Everyday Language. *Risk analysis, 36*(2), 320–338.

Boldrin, M., & Levine, D. K. (2008). *Against Intellectual Monopoly*. Cambridge, UK: Cambridge University Press.

Bolger, F., & Rowe, G. (2015a). The aggregation of expert judgment: Do good things come to those who weight? *Risk Analysis, 35*(1), 5–11. DOI: https://doi.org/10.1111/risa.12272

Bolger, F., & Rowe, G. (2015b). There is Data, and then there is Data: Only Experimental Evidence will Determine the Utility of Differential Weighting of Expert Judgment. *Risk Analysis, 35*(1), 21–6. DOI: https://doi.org/10.1111/risa.12345

Boulton, G. (2012). *Science as an open enterprise.* London: The Royal Society.

Bowles, M. D. (1996). U.S. technological enthusiasm and British technological skepticism in the age of the analog brain. *IEEE Annals of the History of Computing, 18*(4), 5–15. DOI: https://doi.org/10.1109/85.539911

Bozeman, B. (2007). *Public Values and Public Interest: Counterbalancing Economic Individualism.* Washington, DC: Georgetown University Press.

Brännström, M. et al. (2014). Livebirth after uterus transplantation. *Lancet, 385*(9968), 607–616. DOI: https://doi.org/10.1016/S0140-6736(14)61728-1

Brauer, F., Feng, Z., & Castillo-Chavez, C. (2010). Discrete Epidemic Models. *Mathematical Biosciences and Engineering, 7*(1), 1–15. DOI: https://doi.org/10.3934/mbe

Brown, G. E. (1999). Past and Prologue: Why I Am Optimistic About the Future. In *AAAS Science and Technology Policy Yearbook.* Washington, DC: AAAS.

Brun, W. (1992). Cognitive components in risk perception: Natural versus manmade risks. *Journal of Behavioral Decision Making, 5*(2), 117–132. DOI: https://doi.org/10.1002/bdm.3960050204

Bush, V. (1945). Science - the endless frontier. *Transactions of the Kansas Academy of Science, 48*(3), 231–264.

Butler, D. (2014). Biosafety controls come under fire. *Nature, 511*(7511), 515–6. DOI: https://doi.org/10.1038/511515a

Cairney, P. (2016). *The Politics of Evidence-Based Policy Making.* 1st ed. 2016 edition. London: Palgrave Pivot.

Calabrese, E. J., & Baldwin, L. A. (2003). Toxicology rethinks its central belief. *Nature, 421*(6924), 691–2. DOI: https://doi.org/10.1038/421691a

Calow, P. (2014). Environmental Risk Assessors as Honest Brokers or Stealth Advocates. *Risk Analysis, 34*(11), 1972–1977. DOI: https://doi.org/10.1111/risa.12225

Camerer, C., Loewenstein, G., & Weber, M. (1989). The Curse of Knowledge in Economic Settings: An Experimental Analysis. *Journal of Political Economy, 97*(5), 1232–1254.

Cameron, N. M. de S., & Caplan, A. (2009). Our synthetic future. *Nature Biotechnology, 27*, 1103–1105. DOI: https://doi.org/10.1038/nbt1209-1103.

Carson, R. (1962). *Silent Spring.* Boston: Houghton Mifflin.

Carus, W. S. (2015). The History of Biological Weapons Use: What We Know and What We Don't. *Health Security, 13*(4), 219–255. DOI: https://doi.org/10.1089/hs.2014.0092

Carus, W. S. (2017). *A Short History of Biological Warfare: From Pre-History to the 21st Century*. National Defense University Press (Center for the Study of Weapons of Mass Destruction Occasional Paper, No. 12). Available at: http://ndupress.ndu.edu/Portals/68/Documents/occasional/cswmd/CSWMD_OccasionalPaper-12.pdf?ver=2017-08-07-142315-127

Casadevall, A. et al. (2015). Dual-Use Research of Concern (DURC) Review at American Society for Microbiology Journals. *mBio, 6*(4), e01236-15. DOI: https://doi.org/10.1128/mBio.01236-15

Casadevall, A., & Imperiale, M. J. (2014). Risks and Benefits of Gain-of-Function Experiments with Pathogens of Pandemic Potential, Such as Influenza Virus: A Call for a Science-Based Discussion. *mBio, 5*(4), e01730-14. DOI: https://doi.org/10.1128/mBio.01730-14

Casman, E. A., Morgan, M. G., & Dowlatabadi, H. (1999). Mixed Levels of Uncertainty in Complex Policy Models. *Risk Analysis, 19*(1), 33–42.

Castelvecchi, D. (2015). Feuding physicists turn to philosophy for help. *Nature News, 528*(7583), 446. DOI: https://doi.org/10.1038/528446a

CDC. (2014). *Report on the Potential Exposure to Anthrax*. Atlanta, GA: Centers for Disease Control and Prevention.

Cello, J., Paul, A. V., &Wimmer, E. (2002). Chemical synthesis of poliovirus cDNA: Generation of infectious virus in the absence of natural template. *Science, 297*(5583), 1016–8. DOI: https://doi.org/10.1126/science.1072266

Chaiken, S. (1980). Heuristic versus systematic information processing and the use of source versus message cues in persuasion. *Journal of Personality and Social Psychology*, 39(5), 752–766.

Chameau, J.-L., Ballhaus, W. F., and Lin, H. S. (2014). *Emerging and Readily Available Technologies and National Security-A Framework for Addressing Ethical, Legal, and Societal Issues*. Washington, DC: National Academies Press.

Cho, M. K. et al. (1999). Ethical Considerations in Synthesizing a Minimal Genome. *Science, 286*(5447), 2087–2090. DOI: https://doi.org/10.1126/science.286.54 47.2087

Choi, S. (2013). Public Perception and Acceptability of Technological Risk: Policy Implications for Governance. *Journal of Convergence Information Technology, 8*(13), 605–615.

Chorost, M. (2005). *Rebuilt: How Becoming Part Computer Made Me More Human*. New York: Houghton Mifflin.

Chorost, M. (2011). *World Wide Mind: The Coming Integration of Humanity, Machines, and the Internet*. New York: Free Press.

Cirković, M. M. (2012). Small theories and large risks--is risk analysis relevant for epistemology? *Risk analysis, 32*(11), 1994–2004. DOI: https://doi.org/10.1111/j.1539-6924.2012.01914.x

Clahsen, S. C. S. et al. (2019). Why Do Countries Regulate Environmental Health Risks Differently? A Theoretical Perspective. *Risk Analysis, 39*(2), 439–461. DOI: https://doi.org/10.1111/risa.13165

Clark, M. (2013). *Thanks JFK: States Gained from Space Program, Stateline.* Pew Charitable Trusts. Available at: http://www.pewstates.org/projects/stateline/headlines/thanks-jfk-states-gained-from-space-program-85899521645

Clauset, A., Larremore, D. B., & Sinatra, R. (2017). Data-driven predictions in the science of science. *Science, 355*(6324), 477–480. DOI: https://doi.org/10.1126/science.aal4217

Clauset, A., Shalizi, C., & Newman, M. (2009). Power-Law Distributions in Empirical Data. *SIAM Review, 51*(4), 661–703. DOI: https://doi.org/10.1137/070710111

Claxton, K. P., & Sculpher, M. J. (2006). Using Value of Information Analysis to Prioritise Health Research. *PharmacoEconomics, 24*(11), 1055–1068.

Coeckelbergh, M. (2009). Risk and Public Imagination: Mediated Risk Perception as Imaginative Moral Judgement. In L. Asveld & S. Roeser (Eds.), *The Ethics of Technological Risk* (pp. 202–219). London: Routledge.

Cohen, J. (2018). Universal flu vaccine is 'an alchemist's dream'. *Science, 362*(6419), 1094. DOI: https://doi.org/10.1126/science.362.6419.1094

Cohen, W. M., & Levinthal, D. A. (1989). Innovation and learning: The two faces of R&D. *The Economic Journal, 99*(397), 569–596.

Commoner, B. (1971). *The Closing Circle: Nature, Man, and Technology.* New York: Knopf.

Cooke, R. M. (2015). The aggregation of expert judgment: Do good things come to those who weight? *Risk Analysis, 35*(1), 12–5. DOI: https://doi.org/10.1111/risa.12353

Costanza, R. (1999). Four visions of the century ahead: Will it be Star Trek, Ecotopia, Big Government, or Mad Max. *The Futurist, 33*(2), 23–28.

Costanza, R. (2000). Visions of alternative (unpredictable) futures and their use in policy analysis. *Conservation Ecology, 4*(1), 5.

Cox, L. A. (2008). What's wrong with risk matrices? *Risk Analysis, 28*(2), 497–512. DOI: https://doi.org/10.1111/j.1539-6924.2008.01030.x

Cox, L. A. (2012a). Community resilience and decision theory challenges for catastrophic events. *Risk Analysis, 32*(11), 1919–34. DOI: https://doi.org/10.1111/j.1539-6924.2012.01881.x

Cox, L. A. (2012b). Confronting deep uncertainties in risk analysis. *Risk Analysis, 32*(10), 1607–29. DOI: https://doi.org/10.1111/j.1539-6924.2012.01792.x

Cox, L. A., Babayev, D., & Huber, W. (2005). Some limitations of qualitative risk rating systems. *Risk Analysis, 25*(3), 651–62. DOI: https://doi.org/10.1111/j.1539-6924.2005.00615.x

Cranor, C. F. (1997). Normative Nature of Risk Assessment: The Features and Possibilities. *Risk: Health, Safety & Environment, 8*, 123–136.

Cranor, C. F. (2009). A Plea for a Rich Conception of Risks. In L. Asveld & S. Roeser (Eds.), *The Ethics of Technological Risk* (pp. 27–39). London: Routledge.

Crease, R. P. (2019). The rise and fall of scientific authority — and how to bring it back. *Nature, 567*, 309–310. DOI: https://doi.org/10.1038/d41586-019-00872-w

Cropper, M. L., Aydede, S. K., & Portney, P. R. (1994). Preferences for life saving programs: how the public discounts time and age. *Journal of Risk and Uncertainty, 8*(3), 243–265.

Csiszar, A. (2016). Peer review: Troubled from the start. *Nature, 532*(7599), 306. DOI: https://doi.org/10.1038/532306a

Cyranoski, D. (2019). Russian biologist plans more CRISPR-edited babies. *Nature, 570*(7760), 145–146. DOI: https://doi.org/10.1038/d41586-019-01770-x.

Czarnitzki, D., & Lopes-Bento, C. (2013). Value for money? New micro-econometric evidence on public R&D grants in Flanders. *Research Policy, 42*(1), 76–89.

Daar, J., & Klipstein, S. (2016). Refocusing the ethical choices in womb transplantation. *Journal of Law and the Biosciences, 3*(2), 383–388. DOI: https://doi.org/10.1093/jlb/lsw031

Davidson, M. D. (2009). Acceptable Risk to Future Generations. In L. Asveld & S. Roeser (Eds.), *The Ethics of Technological Risk* (pp. 77–91). London: Routledge.

DeLoache, W. C. et al. (2015). An enzyme-coupled biosensor enables (S-reticuline production in yeast from glucose. *Nature Chemical Biology.* DOI: https://doi.org/10.1038/nchembio.1816

Dempster, A. P. (1967). Upper and Lower Probabilities Induced by a Mulitvalued Mapping. *The Annals of Mathematical Statistics, 38*(2), 325–339.

DOJ (2010). *Amerithrax Investigative Summary.* Washington, DC: U.S. Department of Justice.

Donnelly, C. A. et al. (2018). Four principles to make evidence synthesis more useful for policy. *Nature, 558*(7710), 361. DOI: https://doi.org/10.1038/d41586-018-05414-4

Doorn, N., & Hansson, S. O. (2011). Should Probabilistic Design Replace Safety Factors? *Philosophy & Technology, 24*(2), 151–168. DOI: https://doi.org/10.1007/s13347-010-0003-6

Doudna, J. (2015). Genome-editing revolution: My whirlwind year with CRISPR. *Nature, 528*(7583), 469–471. DOI: https://doi.org/10.1038/528469a

Douglas, H. (2000). Inductive Risk and Values in Science. *Philosophy of Science, 67*(4), 559–579. DOI: https://doi.org/10.1086/392855

Douglas, M., & Wildavsky, A. (1982). *Risk and Culture: An Essay on the Selection of Technological and Environmental Dangers.* Berkeley, CA: University of California Press.

Draper, D. (1995). Assessment and Propagation of Model Uncertainty. *Journal of the Royal Statistical Society. Series B (Methodological), 57*(1), 45–97.

Drinkwater, K. et al. (2014) *Creating a Research Agenda for the Ecological Implications of Synthetic Biology.* Washington, DC, 36–36.

Drubin, D. G., & Oster, G. (2010). Experimentalist meets theoretician: A tale of two scientific cultures. *Molecular Biology of the Cell, 21*(13), 2099–2101. DOI: https://doi.org/10.1091/mbc.E10-02-0143

Dubois, D. (2006). Possibility theory and statistical reasoning. *Computational Statistics & Data Analysis, 51*(1), 47–69. DOI: https://doi.org/10.1016/j. csda.2006.04.015

Dubois, D., & Prade, H. (1988). *Possibility Theory: An Approach to Computerized Processing of Uncertainty*. New York: Plenum Press.

Eagly, A. H., & Chaiken, S. (1993). *The psychology of attitudes*. Fort Worth, TX: Harcourt, Brace, & Jovanovich.

Edwards, A. et al. (2001). Presenting risk information – A review of the effects of 'framing' and other manipulations on patient outcomes. *Journal of Health Communication, 6*(1), 61–82. DOI: https://doi.org/10.1080/10810 730150501413

Edwards, D. W. (2005). Are we too Risk-Averse for Inherent Safety? *Process Safety and Environmental Protection, 83*(2), 90–100. DOI: https://doi. org/10.1205/psep.04309

Efron, B., & Tibshirani, R. J. (1993). *An Introduction to the Bootstrap*. Boca Raton, FL: Chapman & Hall/CRC.

Elahi, S. (2011). Here be dragons... exploring the 'unknown unknowns.' *Futures, 43*(2), 196–201. DOI: https://doi.org/10.1016/j.futures.2010.10.008

Ellul, J. (1964). *The Technological Society*. English Tr. Edited by J. Wilkinson. New York: Vintage Books.

Engel-Glatter, S. (2013). Dual-use research and the H5N1 bird flu: Is restricting publication the solution to biosecurity issues? *Science and Public Policy, 41*(3), 370–383. DOI: https://doi.org/10.1093/scipol/sct064

Ernst, E., & Resch, K. L. (1996). Risk-benefit ratio or risk-benefit nonsense? *Journal of clinical epidemiology, 49*(10), 1203–4.

Espinoza, N. (2009) Incommensurability: The Failure to Compare Risks. In L. Asveld & S. Roeser (Eds.), *The Ethics of Technological Risk* (pp. 128–143). London: Routledge.

Esvelt, K. M. et al. (2014). Emerging Technology: Concerning RNA-guided gene drives for the alteration of wild populations. *eLife, 3*, e03401. DOI: https://doi.org/10.7554/eLife.03401

Esvelt, K. M., & Gemmell, N. J. (2017). Conservation demands safe gene drive. *PLOS Biology, 15*(11), e2003850. DOI: https://doi.org/10.1371/journal. pbio.2003850

Evans, J. H. (2018). *Morals Not Knowledge: Recasting the Contemporary U.S. Conflict Between Religion and Science*. University of California Press. DOI: https://doi.org/https://doi.org/10.1525/luminos.47

Evans, N. G. (2013). Great expectations - ethics, avian flu and the value of progress. *Journal of medical ethics, 39*(4), 209–13. DOI: https://doi.org/10.1136/ medethics-2012-100712

Eypasch, E. et al. (1995). Probability of adverse events that have not yet occurred: A statistical reminder. *BMJ, 311*(7005), 619–620. DOI: https:// doi.org/10.1136/bmj.311.7005.619

Falk, A., & Szech, N. (2013). Morals and Markets. *Science, 340*(6133), 707–711. DOI: https://doi.org/10.1126/science.1231566

Fang, J. (2010). Ecology: A world without mosquitoes. *Nature, 466*(7305), 432–4. DOI: https://doi.org/10.1038/466432a

Farquhar, S., Cotton-Barratt, O., & Snyder-Beattie, A. (2017). Pricing Externalities to Balance Public Risks and Benefits of Research. *Health Security, 15*(4), 401–408. DOI: https://doi.org/10.1089/hs.2016.0118

Farrell, R. M., & Falcone, T. (2015). Uterine transplant: new medical and ethical considerations. *Lancet, 385*(9968), 581–2. DOI: https://doi.org/10.1016/S0140-6736(14)61792-X

Fears, R., & ter Meulen, V. (2016). European Academies Advise on Gain-of-Function Studies in Influenza Virus Research. *Journal of Virology, 90*(5), 2162–2164. DOI: https://doi.org/10.1128/JVI.03045-15

Federal Demonstration Partnership. (2013). *STAR METRICS – Phase II, National Academy of Sciences*. Available at: http://sites.nationalacademies.org/PGA/fdp/PGA_057159

Ferson, S. (2014). *Model uncertainty in risk analysis*. Compiègne Cedex, France: Université de Technologie de Compiègne.

Ferson, S., & Ginzburg, L. R. (1996). Different methods are needed to propagate ignorance and variability. *Reliability Engineering & System Safety, 54*(2–3), 133–144. DOI: https://doi.org/10.1016/S0951-8320(96)00071-3

Ferson, S., & Hajagos, J. G. (2004). Arithmetic with uncertain numbers: rigorous and (often) best possible answers. *Reliability Engineering & System Safety, 85*(1–3), 135–152. DOI: https://doi.org/10.1016/j.ress.2004.03.008

Festinger, L. (1957). *A Theory of Cognitive Dissonance*. Redwood City, CA: Stanford University Press.

Feyerabend, P. (1970). Consolations for the specialist. In I. Lakatos & A. Musgrave (Eds.), *Criticism and the Growth of Knowledge* (pp. 197–230). Cambridge, UK: Cambridge University Press.

Feyerabend, P. (1975). *Against Method*. London: New Left Books.

Feyerabend, P. (1978). *Science in a Free Society*. London: New Left Books.

Feyerabend, P. (2011). *The Tyranny of Science*. Edited by E. Oberheim. Cambridge, UK: Polity Press.

Field Jr., T. G. (1994). *The Science Court Symposia, Risk: Health, Safety & Environment*. Available at: http://ipmall.info/risk/scict.htm (Accessed: 6 May 2015).

Fineberg, H. V. (2009). Swine flu of 1976: Lessons from the past. An interview with Dr Harvey V Fineberg. *Bulletin of the World Health Organization, 87*(6), 414–5.

Fineberg, H. V. (2015). Wider attention for GOF science. *Science, 347*(6225), 929.

Fink, G. R. (2004). *Biotechnology Research in an Age of Terrorism*. Washington, DC: National Academies Press.

Fischhoff, B. et al. (1978). How safe is safe enough? A psychometric study of attitudes towards technological risks and benefits. *Policy sciences, 9*(2), 127–152.

Fischhoff, B. (1995). Risk perception and communication unplugged: Twenty years of process. *Risk analysis, 15*(2), 137–45.

Fischhoff, B. (2000). Scientific management of science? *Policy Sciences, 33*(1), 73–87.

Fischhoff, B. (2015). The realities of risk-cost-benefit analysis. *Science, 350* (6260), 527, aaa6516-1-aaa6516-7. DOI: https://doi.org/10.1126/science. aaa6516

Fiske, S. T., & Taylor, S. E. (1984). *Social Cognition.* n edition. Reading, Mass: Longman Higher Education (Topics in Social Psychology).

Flage, R., et al. (2014). Concerns, challenges, and directions of development for the issue of representing uncertainty in risk assessment. *Risk analysis, 34*(7), 1196–207. DOI: https://doi.org/10.1111/risa.12247

Flyvbjerg, B. (2006). Five Misunderstandings About Case-Study Research. *Qualitative Inquiry, 12*(2), 219–245.

Fontana, R. et al. (2013). Reassessing patent propensity: Evidence from a dataset of R&D awards, 1977–2004. *Research Policy.* (Economics, innovation and history: Perspectives in honour of Nick von Tunzelmann), *42*(10), 1780–1792. DOI: https://doi.org/10.1016/j.respol.2012.05.014

Fossati, E. et al. (2015). Synthesis of Morphinan Alkaloids in Saccharomyces cerevisiae. *PloS ONE, 10*(4), e0124459. DOI: https://doi.org/10.1371/journal. pone.0124459

Fouchier, R. A. M. (2015a). Reply to 'Comments on Fouchier's Calculation of Risk and Elapsed Time for Escape of a Laboratory-Acquired Infection from His Laboratory.' *mBio, 6*(2), e00407–15. DOI: https://doi.org/10.1128/ mBio.00407-15

Fouchier, R. A. M. (2015b). Studies on influenza virus transmission between ferrets: The public health risks revisited. *mBio, 6*(1), e02560-14. DOI: https://doi.org/10.1128/mBio.02560-14

Fox, C. R., & Tversky, A. (1995). Ambiguity Aversion and Comparative Ignorance. *The Quarterly Journal of Economics, 110*(3), 585–603.

Fox, D. M. (1995). *Engines of Culture: Philanthropy and Art Museums.* 2nd ed. New Brunswick: Transaction Publishers.

Fox, N. (2002). *Against the Machine: The Hidden Luddite Tradition in Literature, Art, and Individual Lives.* Washington, DC: Island Press.

Franco, A., Malhotra, N., & Simonovits, G. (2014). Publication bias in the social sciences: Unlocking the file drawer. *Science, 345*(6203), 1502–1505. DOI: https://doi.org/10.1126/science.1255484

Friedkin, N. E. et al. (2016). Network science on belief system dynamics under logic constraints. *Science, 354*(6310), 321.–326. DOI: https://doi. org/10.1126/science.aag2624

Fuller, S., & Lipinska, V. (2014). *The Proactionary Imperative: A Foundation for Transhumanism.* Basingstoke, UK: Palgrave Macmillan.

Funk, C., Rainie, L., & Page, D. (2015). *Americans, Politics and Science Issues.* Pew Research Center.

Furmanski, M. (2015). The 1977 H1N1 Influenza Virus Reemergence Demonstrated Gain-of-Function Hazards. *mBio, 6*(5), e01434-15. DOI: https://doi. org/10.1128/mBio.01434-15

Galanie, S. et al. (2015). Complete biosynthesis of opioids in yeast. *Science.* DOI: https://doi.org/10.1126/science.aac9373

Galizi, R. et al. (2014). A synthetic sex ratio distortion system for the control of the human malaria mosquito. *Nature communications, 5,* 3977. DOI: https://doi.org/10.1038/ncomms4977

Gallagher, K. M., & Updegraff, J. A. (2012). Health Message Framing Effects on Attitudes, Intentions, and Behavior: A Meta-analytic Review. *Annals of Behavioral Medicine, 43*(1), 101–116.

Gantz, V. M. et al. (2015). Highly efficient Cas9-mediated gene drive for population modification of the malaria vector mosquito Anopheles stephensi. *Proceedings of the National Academy of Sciences, 112*(49), E6736–E6743. DOI: https://doi.org/10.1073/pnas.1521077112

Gantz, V. M., & Bier, E. (2015). The mutagenic chain reaction: A method for converting heterozygous to homozygous mutations. *Science, 348*(6233), 442–444. DOI: https://doi.org/10.1126/science.aaa5945

Gardoni, P., & Murphy, C. (2014). A scale of risk. *Risk analysis, 34*(7), 1208–27. DOI: https://doi.org/10.1111/risa.12150

Geiser, K. (2015). *Chemicals without Harm.* Cambridge, MA: MIT Press.

Gillette, C. P., & Krier, J. E. (1990). Risk, Courts, and Agencies. *University of Pennsylvania Law Review, 138*(4), 1027–1109.

Glass, K., Xia, Y., & Grenfell, B. T. (2003). Interpreting time-series analyses for continuous-time biological models—measles as a case study. *Journal of Theoretical Biology, 223*(1), 19–25. DOI: https://doi.org/10.1016/S0022-5193(03)00031-6

Glenn, J. (2012). The radium age. *Nature, 489*(7415), 204–205.

Goodwin, P., & Wright, G. (2010). The limits of forecasting methods in anticipating rare events. *Technological Forecasting and Social Change, 77*(3), 355–368. DOI: https://doi.org/10.1016/j.techfore.2009.10.008

Graham, C., Laffan, K., & Pinto, S. (2018). Well-being in metrics and policy. *Science, 362*(6412), 287–288. DOI: https://doi.org/10.1126/science.aau5234

Green, B. P. (2016). Emerging technologies, catastrophic risks, and ethics: three strategies for reducing risk. In *2016 IEEE International Symposium on Ethics in Engineering, Science and Technology (ETHICS). 2016 IEEE International Symposium on Ethics in Engineering, Science and Technology (ETHICS)*, Vancouver, BC: IEEE, 1–7. DOI: https://doi.org/10.1109/ETHICS.2016.7560046.

Griliches, Z. (1979). Issues in assessing the contribution of research and development to productivity growth. *Bell Journal of Economics, 10,* 92–116.

Griliches, Z. (1994). Productivity, R&D, and the data constraint. *The American Economic Review, 84*(1), 1–23.

Gronvall, G. K. (2015). US Competitiveness in Synthetic Biology. *Health Security, 13*(6), 378–389. DOI: https://doi.org/10.1089/hs.2015.0046

Gross, L. (2009). A broken trust: Lessons from the vaccine--autism wars. *PLoS biology, 7*(5), e1000114.

Gruner, S. (2008). New Ethics of Technology: A Discourse in Germany in the 1990s. In *Annual Conference of the Philosophical Society of Southern Africa.* Pretoria, p. 15.

Grunwald, H. A. et al. (2019). Super-Mendelian inheritance mediated by CRISPR–Cas9 in the female mouse germline. *Nature, 566*(7742), 105–109. DOI: https://doi.org/10.1038/s41586-019-0875-2

Gryphon Scientific. (2015). *Risk and Benefit Analysis of Gain of Function Research. Draft Final Report.* Takoma Park, MD.

Guston, D. H. (2000). *Between Politics and Science: Assuring the Integrity and Productivity of Reseach.* 1st US-1st Printing edition. Cambridge, NY: Cambridge University Press.

Guston, D. H. et al. (2014). Responsible innovation: Motivations for a new journal. *Journal of Responsible Innovation, 1*(1), 1–8. DOI: https://doi.org/1 0.1080/23299460.2014.885175

Guston, D. H., & **Sarewitz, D.** (2002). Real-time technology assessment. *Technology in Society.* (American Perspectives on Science and Technology Policy), *24*(1), 93–109. DOI: https://doi.org/10.1016/S0160-791X(01)00047-1

Haidt, J. (2012). *The Righteous Mind: Why Good People are Divided by Politics and Religion.* New York: Pantheon Books.

Haimes, Y. Y. (1981). Hierarchical Holographic Modeling. *IEEE Transactions on Systems, Man, and Cybernetic, 11*(9), 606–617. DOI: https://doi.org/10.1109/ TSMC.1981.4308759

Haimes, Y. Y., Kaplan, S., & **Lambert, J. H.** (2002). Risk Filtering, Ranking, and Management Framework Using Hierarchical Holographic Modeling. *Risk Analysis, 22*(2), 383–397. DOI: https://doi.org/10.1111/0272-4332.00020

Hall, B. H., Mairesse, J., & **Mohnen, P.** (2009). *Measuring the returns to R&D.* 15622. Cambridge, MA: National Bureau of Economic Research.

Hall, C. A. S. (1988). An assessment of several of the historically most influential theoretical models used in ecology and of the data provided in their support. *Ecological Modelling, 43*(1–2), 5–31. DOI: https://doi.org/10.1016/0304-3800(88)90070-1

Hammitt, J. K., & **Zhang, Y.** (2013). Combining experts' judgments: Comparison of algorithmic methods using synthetic data. *Risk analysis, 33*(1), 109–20. DOI: https://doi.org/10.1111/j.1539-6924.2012.01833.x

Hammond, A. et al. (2016). A CRISPR-Cas9 gene drive system targeting female reproduction in the malaria mosquito vector *Anopheles gambiae. Nature Biotechnology, 34*(1), 78–83. DOI: https://doi.org/10.1038/nbt.3439

Hanea, A. M. et al. (2018). The Value of Performance Weights and Discussion in Aggregated Expert Judgments. *Risk Analysis, 38*(9), 1781–1794. DOI: https://doi.org/10.1111/risa.12992

Hansson, S. O. (1999). The Moral Significance of Indetectable Effects. *Risk: Health, Safety & Environment, 10,* 101–108.

Hansson, S. O. (2003). Are natural risks less dangerous than technological risks? *Philosophia Naturalis, 40,* 43–54.

Hansson, S. O. (2005). Seven Myths of Risk. *Risk Management, 7*(2), 7–17.

Hansson, S. O. (2009). From the casino to the jungle. *Synthese, 168*(3), 423–432. DOI: https://doi.org/10.1007/s11229-008-9444-1

Hansson, S. O. (2010a). Promoting inherent safety. *Process Safety and Environmental Protection, 88*(3), 168–172. DOI: https://doi.org/10.1016/j.psep.2010.02.003

Hansson, S. O. (2010b). Risk: Objective or subjective, facts or values. *Journal of Risk Research, 13*(2), 231–238. DOI: https://doi.org/10.1080/13669870903126226

Hansson, S. O. (2012). A Panorama of the Philosophy of Risk. In S. Roeser et al. (Eds.), *Handbook of Risk Theory* (ppp. 27–54). Dordrecht: Springer Science & Business Media.

Hansson, S. O. (2013). *The Ethics of Risk: Ethical Analysis in an Uncertain World.* 2013 edition. Basingstoke, UK ; New York, NY: Palgrave Macmillan.

Hansson, S. O. (2018). How to Perform an Ethical Risk Analysis (eRA). *Risk Analysis, 38*(9), 1820–1829. DOI: https://doi.org/10.1111/risa.12978

Hansson, S. O., & Aven, T. (2014). Is risk analysis scientific? *Risk analysis, 34*(7), 1173–83. DOI: https://doi.org/10.1111/risa.12230

Hayes, K. R. et al. (2015). *Risk Assessment for Controlling Mosquito Vectors with Engineered Nucleases: Sterile Male Construct Final report.* Hobart, Australia: CSIRO Biosecurity Flagship.

Healy, K. (2017). Fuck Nuance. *Sociological Theory, 35*(2), 118–127. DOI: https://doi.org/10.1177/0735275117709046

Heidegger, M. (1977). *The Question Concerning Technology and Other Essays.* English Tr. Edited by W. Lovitt. New York: Garland.

Hempel, C. G. (1960). Inductive Inconsistencies. *Synthese, 12*(4), 439–469.

Henderson, D. A., & Arita, I. (2014). The smallpox threat: A time to reconsider global policy. *Biosecurity and bioterrorism, 12*(3), 1–5. DOI: https://doi.org/10.1089/bsp.2014.1509.comm

Henkel, R. D., Miller, T., & Weyant, R. S. (2012). Monitoring Select Agent Theft, Loss and Release Reports in the United States — 2004-2010. *Applied Biosafety, 17*(4), 171–180.

Henry, H., & Taylor, A. (2009). Re-thinking Apollo: Envisioning environmentalism in space. *The Sociological Review, 57*(s1), 190–203.

Herfst, S. et al. (2012). Airborne transmission of influenza A/H5N1 virus between ferrets. *Science, 336*(6088), 1534–41. DOI: https://doi.org/10.1126/science.1213362

Hergarten, S. (2004). Aspects of risk assessment in power-law distributed natural hazards. *Nat. Hazards Earth Syst. Sci., 4*(2), 309–313. DOI: https://doi.org/10.5194/nhess-4-309-2004

Hermansson, H., & Hansson, S. O. (2007). A Three-Party Model Tool for Ethical Risk Analysis. *Risk Management, 9*(3), 129–144.

Hites, R. A. et al. (2004). Global assessment of organic contaminants in farmed salmon. *Science, 303*(5655), 226–9. DOI: https://doi.org/10.1126/science.1091447

Hoeting, J. A. et al. (1999). Bayesian model averaging: A tutorial (with comments by M. Clyde, David Draper and E. I. George, and a rejoinder by the authors). *Statistical Science, 14*(4), 382–417.

Holden, W. L. (2003). Benefit-risk analysis: A brief review and proposed quantitative approaches. *Drug Safety, 26*(12), 853–862. DOI: https://doi.org/10.2165/00002018-200326120-00002

Hollanders, H., & Es-Sadki, N. (2014). *Innovation Union Scoreboard 2014, European Commission, UNU-MERIT.* European Commission, UNU-MERIT.

Hossenfelder, S. (2018). *Lost in Math: How Beauty Leads Physics Astray.* New York: Basic Books.

Howard, R. A. (1980). On Making Life and Death Decisions. In R. C. Schwing & W. A. Albers Jr (Eds.), *Societal Risk Assessment: How Safe is Safe Enough?* (pp. 90–113) New York: Plenum Press.

Howard, R. A., & Matheson, J. E. (2005). Influence Diagram Retrospective. *Decision Analysis, 2*(3), 144–147.

Hutchison, C. A. et al. (2016). Design and synthesis of a minimal bacterial genome. *Science, 351*(6280), aad6253. DOI: https://doi.org/10.1126/science.aad6253

Imai, M. et al. (2012). Experimental adaptation of an influenza H5 HA confers respiratory droplet transmission to a reassortant H5 HA/H1N1 virus in ferrets. *Nature, 486*(7403), 420–8. DOI: https://doi.org/10.1038/nature10831

Impagliazzo, A. et al. (2015). A stable trimeric influenza hemagglutinin stem as a broadly protective immunogen. *Science, 349*(6254), 1301–1306. DOI: https://doi.org/10.1126/science.aac7263

Imperiale, M. J., & Casadevall, A. (2016). Zika Virus Focuses the Gain-of-Function Debate. *mSphere, 1*(2), e00069-16. DOI: https://doi.org/10.1128/mSphere.00069-16

Inglesby, T. V., & Relman, D. A. (2016). How likely is it that biological agents will be used deliberately to cause widespread harm? *EMBO Reports, 17*(2), 127–130. DOI: https://doi.org/10.15252/embr.201541674

Institute of Medicine. (2009). *On Being a Scientist: A Guide to Responsible Conduct in Research: Third Edition.* Washington, DC: National Academies Press. DOI: https://doi.org/10.17226/12192

Ioannidis, J. P. A. (2005). Why most published research findings are false. *PLoS medicine, 2*(8), e124. DOI: https://doi.org/10.1371/journal.pmed.0020124

Jaczko, G. B. (2019). *Confessions of a Rogue Nuclear Regulator.* New York: Simon & Schuster.

Jaffe, A. B., Trajtenberg, M., & Henderson, R. (1993). Geographic Localization of Knowledge Spillovers as Evidence by Patent Citations. *Quarterly Journal of Economics, 108*(3), 577–598.

Jasanoff, S. (1993). Bridging the Two Cultures of Risk Analysis. *Risk Analysis, 13*(2), 123–129. DOI: https://doi.org/10.1111/j.1539-6924.1993.tb01057.x

Jasanoff, S. (2003). Technologies of humility: Citizen participation in governing science. *Minerva, 41*(3), 223–244. DOI: https://doi.org/10.2307/41821248

Jasanoff, S. (2004). *States of Knowledge: The Co-production of Science and Social Order.* London: Routledge.

Jasanoff, S. (2007). Technologies of humility. *Nature, 450*(7166), 33. DOI: https://doi.org/10.1038/450033a

Jasanoff, S. (2016). *The Ethics of Invention: Technology and the Human Future.* New York: W. W. Norton & Company.

Jasanoff, S., & Hurlbut, J. B. (2018). A global observatory for gene editing. *Nature, 555*(7697), 435–437. DOI: https://doi.org/10.1038/d41586-018-03270-w

Johnson, B. D. (2011). *Science Fiction Prototyping: Designing the Future with Science Fiction.* 1 edition. San Rafael, Calif.: Morgan & Claypool Publishers.

Johnson, M. (1993). *Moral Imagination: Implications of Cognitive Science for Ethics.* Chicago: University of Chicago Press.

Johnson, R. W. (2000). Analyze hazards, not just risks. *Chemical Engineering Progress, 96*(7), 31–40.

Jonas, H. (1984). *The Imperative of Responsibility: In Search of an Ethics for the Technological Age.* Chicago: University of Chicago Press.

Jones, S. E. (2006). *Against Technology: From the Luddites to Neo-Luddism.* New York: Routledge.

Joy, B. (2000, April 8). Why the future doesn't need us. *Wired.*

Kaebnick, G. E. et al. (2016). Precaution and governance of emerging technologies. *Science, 354*(6313), 710–711. DOI: https://doi.org/10.1126/science.aah5125

Kahan, D. et al. (2006). Fear of Democracy: a Cultural Evaluation of Sunstein on Risk. *Harvard Law Review, 119*(4), 1071–1109. DOI: https://doi.org/10.2139/ssrn.801964

Kahan, D. (2010). Fixing the communications failure. *Nature, 463*(7279), 296–7. DOI: https://doi.org/10.1038/463296a

Kahan, D. et al. (2010). Who fears the HPV vaccine, who doesn't, and why? An experimental study of the mechanisms of cultural cognition. *Law and Human Behavior, 34*(6), 501–516. DOI: https://doi.org/10.1007/s10979-009-9201-0

Kahan, D. *et al.* (2015). Geoengineering and Climate Change Polarization: Testing a Two-Channel Model of Science Communication. *The Annals of the American Academy of Political and Social Science, 658*(1), 192–222. DOI: https://doi.org/10.1177/0002716214559002

Kahan, D., Jenkins-Smith, H., & Braman, D. (2011). Cultural cognition of scientific consensus. *Journal of Risk Research, 14*(2), 147–174. DOI: https://doi.org/10.1080/13669877.2010.511246

Kahlor, L. et al. (2003). Studying Heuristic-Systematic Processing of Risk Communication. *Risk Analysis, 23*(2), 355–368. DOI: https://doi.org/10.1111/1539-6924.00314

Kahn, L. H. (2012). Can biosecurity be embedded into the culture of the life sciences? *Biosecurity and bioterrorism, 10*(2), 241–6. DOI: https://doi.org/10.1089/bsp.2012.0023

Kahneman, D. (2011) *Thinking, Fast and Slow.* 1 edition. New York: Farrar, Straus and Giroux.

Kahneman, D., & **Tversky, A.** (1979). Prospect Theory: An Analysis of Decision under Risk. *Econometrica: Journal of the Econometric Society, 47*(3), 263–291. DOI: https://doi.org/10.1111/j.1536-7150.2011.00774.x

Kaiser, J. (2019). Controversial flu studies can resume, U.S. panel says. *Science, 363*(6428), 676–677. DOI: https://doi.org/10.1126/science.363.6428.676

Kandel, E. (2016). *Reductionism in Art and Brain Science: Bridging the Two Cultures.* 1 edition. New York: Columbia University Press.

Kantrowitz, A. (1967). Proposal for an Institution for Scientific Judgment. *Science, 156*(3776), 763–764.

Kasperson, R. E. et al. (1988). The Social Amplification of Risk: A Conceptual Framework. *Risk Analysis, 8*(2), 177–187. DOI: https://doi.org/10.1111/j.1539-6924.1988.tb01168.x

Kass, L. R. (1997, June 2). The Wisdom of Repugnance. *The New Republic.*

Kass, L. R. (2009). Forbidding Science: Some Beginning Reflections. *Science and Engineering Ethics, 15*(3), 271–282. DOI: https://doi.org/10.1007/s11948-009-9122-9

Ke, Q. et al. (2015). Defining and identifying Sleeping Beauties in science. *Proceedings of the National Academy of Sciences,* 1–6.

Keisler, J. M. *et al.* (2013). Value of information analysis: The state of application. *Environment Systems and Decisions, 34*(1), 3–23.

Kelle, A. (2013). Beyond Patchwork Precaution in the Dual-Use Governance of Synthetic Biology. *Science and Engineering Ethics, 19*(3), 1121–1139. DOI: https://doi.org/10.1007/s11948-012-9365-8

Kelly, K. (2010). *What Technology Wants.* New York: Viking Press.

Keulartz, J., & **van den Belt, H.** (2016). DIY-Bio – economic, epistemological and ethical implications and ambivalences. *Life Sciences, Society and Policy, 12*(1), 7. DOI: https://doi.org/10.1186/s40504-016-0039-1

Khan, F. I., & **Amyotte, P. R.** (2003). How to Make Inherent Safety Practice a Reality. *The Canadian Journal of Chemical Engineering, 81*(1), 2–16. DOI: https://doi.org/10.1002/cjce.5450810101

Kidd, D. C., & **Castano, E.** (2013). Reading literary fiction improves theory of mind. *Science, 342*(6156), 377–80. DOI: https://doi.org/10.1126/science.1239918

Kim, J. *et al.* (2014). Disentangling the influence of value predispositions and risk/benefit perceptions on support for nanotechnology among the American public. *Risk analysis, 34*(5), 965–80. DOI: https://doi.org/10.1111/risa.12141

Kiran, A. H., & **Verbeek, P.-P.** (2010). Trusting Our Selves to Technology. *Knowledge, Technology & Policy, 23*(3–4), 409–427. DOI: https://doi.org/10.1007/s12130-010-9123-7

Kitcher, P. (2001). *Science, Truth, and Democracy*. Oxford: Oxford University Press.

Kletz, T. A. (1978). What you don't have, can't leak. *Chemistry and Industry, 6*, 287–292.

Kletz, T. A. (1985). Inherently safer plants. *Plant/Operations Progress, 4*(3), 164–167. DOI: https://doi.org/10.1002/prsb.720040311

Klotz, L. C. (2015). Comments on Fouchier's calculation of risk and elapsed time for escape of a laboratory-acquired infection from his laboratory. *mBio, 6*(2), e00268-15. DOI: https://doi.org/10.1128/mBio.00268-15

Klotz, L. C., & Sylvester, E. J. (2009). *Breeding Bio Insecurity*. Chicago: University of Chicago Press.

Klotz, L. C., & Sylvester, E. J. (2012). *The unacceptable risks of a man-made pandemic, Bulletin of the Atomic Scientists*. Available at: http://thebulletin.org/unacceptable-risks-man-made-pandemic.

Klotz, L. C., & Sylvester, E. J. (2014). The Consequences of a Lab Escape of a Potential Pandemic Pathogen. *Frontiers in Public Health, 2*, 116. DOI: https://doi.org/10.3389/fpubh.2014.00116

Koblentz, G. D. (2017). The De Novo Synthesis of Horsepox Virus: Implications for Biosecurity and Recommendations for Preventing the Reemergence of Smallpox. *Health Security, 15*(6), 620–628. DOI: https://doi.org/10.1089/hs.2017.0061

Koblentz, G. D. (2018). A Critical Analysis of the Scientific and Commercial Rationales for the De Novo Synthesis of Horsepox Virus. *mSphere, 3*(2). DOI: https://doi.org/10.1128/mSphere.00040-18

Kranzberg, M. (1967). The Unity of Science-Technology. *American Scientist, 55*(1), 48–66.

Kranzberg, M. (1968). The Disunity of Science-Technology. *American Scientist, 56*(1), 21–34.

Kuhn, T. S. (1962). *The structure of scientific revolutions*. Chicago: University of Chicago Press.

Kun, H. (2015). Moving away from metrics. *Nature, 520*(7549), S18-20.

Kuzma, J., & Besley, J. C. (2008). Ethics of risk analysis and regulatory review: From bio- to nanotechnology. *NanoEthics, 2*(2), 149–162. DOI: https://doi.org/10.1007/s11569-008-0035-x

Kyrou, K. et al. (2018). A CRISPR–Cas9 gene drive targeting *doublesex* causes complete population suppression in caged *Anopheles gambiae* mosquitoes. *Nature Biotechnology, 36*(11), 1062–1066. DOI: https://doi.org/10.1038/nbt.4245

Lach, S., & Schankerman, M. (1989). Dynamics of R&D and investment in the scientific sector. *The Journal of Political Economy, 97*(4), 880–904.

Lajoie, M. J. et al. (2013). Genomically Recoded Organisms Expand Biological Functions. *Science, 342*(6156), 357–360. DOI: https://doi.org/10.1126/science.1241459

Lakatos, I. (1970). Falsification and the methodology of scientific research programmes. In I. Lakatos & A. Musgrave (Eds.), *Criticism and the Growth of Knowledge* (pp. 91–196). Cambridge, UK: Cambridge University Press.

Lambert, J. H. et al. (1994). Selection of Probability Distributions in Characterizing Risk of Extreme Events. *Risk Analysis, 14*(5), 731–742.

Lander, E. S. et al. (2019). Adopt a moratorium on heritable genome editing. *Nature, 567*(7747), 165. DOI: https://doi.org/10.1038/d41586-019-00726-5

Lane, J. (2009). Assessing the impact of science funding. *Science, 324*(5932), 1273–1275.

Lane, J., & Bertuzzi, S. (2011). Measuring the Results of Science Investments. *Science, 331*(6018), 678–680.

Largent, M. A., & Lane, J. I. (2012). STAR METRICS and the Science of Science Policy. *Review of Policy Research, 29*(3), 431–438.

Laski, H. J. (1931). *The limitations of the expert.* London: Fabian Society (Fabian Tract, 235). Available at: https://digital.library.lse.ac.uk/objects/lse:wal303heb (Accessed: 16 April 2018).

Lavery, J. V. (2018). Building an evidence base for stakeholder engagement. *Science, 361*(6402), 554–556. DOI: https://doi.org/10.1126/science.aat8429

Ledford, H. (2019). CRISPR babies: When will the world be ready? *Nature, 570*(7761), 293–296. DOI: https://doi.org/10.1038/d41586-019-01906-z

Lee, S.-W. et al. (2012). Attenuated vaccines can recombine to form virulent field viruses. *Science, 337*(6091), 188. DOI: https://doi.org/10.1126/science.1217134

Leitenberg, M., Zilinskas, R. A., & Kuhn, J. H. (2012). *The Soviet Biological Weapons Program: A History.* Cambridge, MA: Harvard Univsersity Press.

Leopold, A. (1949). *A Sand County Almanac: And Sketches Here and There.* Oxford University Press.

Lewis, G. et al. (2019). Information Hazards in Biotechnology. *Risk Analysis, 39*(5), 975–981. DOI: https://doi.org/10.1111/risa.13235

Li, D., & Agha, L. (2015). Big names or big ideas: Do peer-review panels select the best science proposals? *Science, 348*(6233), 434–438.

Li, D., Azoulay, P., & Sampat, B. N. (2017). The applied value of public investments in biomedical research. *Science, 356*(6333), 78–81. DOI: https://doi.org/10.1126/science.aal0010

Li, F. C. et al. (2008). Finding the real case-fatality rate of H5N1 avian influenza. *Journal of Epidemiology and Community Health, 62*(6), 555–559. DOI: https://doi.org/10.1136/jech.2007.064030

Liang, P. et al. (2015). CRISPR/Cas9-mediated gene editing in human tripronuclear zygotes. *Protein & cell, 6*(5), 363–72. DOI: https://doi.org/10.1007/s13238-015-0153-5

Liang, P. et al. (2017). Correction of β-thalassemia mutant by base editor in human embryos. *Protein & Cell, 8*(11), 811–822. DOI: https://doi.org/10.1007/s13238-017-0475-6

Linkov, I. et al. (2014). Changing the resilience paradigm. *Nature Climate Change, 4*(6), 407–409. DOI: https://doi.org/10.1038/nclimate2227

Linster, M. et al. (2014). Identification, characterization, and natural selection of mutations driving airborne transmission of A/H5N1 virus. *Cell, 157*(2), 329–39. DOI: https://doi.org/10.1016/j.cell.2014.02.040

Lipsitch, M. (2018). Why Do Exceptionally Dangerous Gain-of-Function Experiments in Influenza? In Y/ Yamauchi (Ed.), *Influenza Virus: Methods and Protocols* (pp. 589–608). New York, NY: Springer New York (Methods in Molecular Biology). DOI: https://doi.org/10.1007/978-1-4939-8678-1_29

Lipsitch, M., & Galvani, A. P. (2014). Ethical alternatives to experiments with novel potential pandemic pathogens. *PLoS medicine, 11*(5), e1001646. DOI: https://doi.org/10.1371/journal.pmed.1001646

Lipsitch, M., & Inglesby, T. V. (2014). Moratorium on Research Intended To Create Novel Potential Pandemic Pathogens. *mBio, 5*(6), e02366-14. DOI: https://doi.org/10.1128/mBio.02366-14

Lipsitch, M., & Inglesby, T. V. (2015). Reply to ' Studies on Influenza Virus Transmission between Ferrets: The Public Health Risks Revisited.' *mBio, 6*(1), e00041-15. DOI: https://doi.org/10.1128/mBio.00041-15

Lipsitch, M. et al. (2016). Viral factors in influenza pandemic risk assessment. *eLife, 5*(NOVEMBER 2016), e18491. DOI: https://doi.org/10.7554/eLife.18491

Loewenstein, G. F. et al. (2001). Risk as feelings. *Psychological bulletin, 127*(2), 267–86.

Loewenstein, G., Moore, D. A., & Weber, R. A. (2003). Paying $1 to lose $2: Misperceptions of the value of information in predicting the performance of others. *Academy of Management Proceedings*, 2003(1), C1–C6.

Lowrie, H., & Tait, J. (2010). *Policy Brief: Guidelines for the Appropriate Risk Governance of Synthetic Biology*. 978-2-9700672-6–9. Geneva: International Risk Governance Council, 52–52. Available at: http://irgc.org/wp-content/uploads/2012/04/irgc_SB_final_07jan_web.pdf.

Lund, E. et al. (2004), Cancer risk and salmon intake. *Science, 305*(5683), 477–8; author reply 477-8. DOI: https://doi.org/10.1126/science.305.5683.477.

Lundberg, A. S., & Novak, R. (2015). CRISPR-Cas Gene Editing to Cure Serious Diseases: Treat the Patient, Not the Germ Line. *The American Journal of Bioethics, 15*(12), 38–40. DOI: https://doi.org/10.1080/15265161.2015.1103817

Ma, H. et al. (2017). Correction of a pathogenic gene mutation in human embryos. *Nature, 548*(7668), 413–419. DOI: https://doi.org/10.1038/nature23305

MacGillivray, B. H. (2014). Heuristics structure and pervade formal risk assessment. *Risk analysis, 34*(4), 771–87. DOI: https://doi.org/10.1111/risa.12136

MacKenzie, C. A. (2014). Summarizing risk using risk measures and risk indices. *Risk analysis, 34*(12), 2143–62. DOI: https://doi.org/10.1111/risa.12220

MacLean, D. (1986). *Values at Risk*. Totawa, NJ: Rowman & Littlefield.

MacLean, D. (2009). Ethics, Reasons and Risk Analysis. In L. Asveld & S. Roeser (Eds.), *The Ethics of Technological Risk* (pp. 115–143). London: Routledge.

Mahmoud, A. (2013). Gain-of-function research: Unproven technique. *Science, 342*(6156), 310–1. DOI: https://doi.org/10.1126/science.342.6156.310-b

Malyshev, D. et al. (2014). A semi-synthetic organism with an expanded genetic alphabet. *Nature, 509*(7500), 385–8. DOI: https://doi.org/10.1038/nature13314

Mandell, D. J. et al. (2015). Biocontainment of genetically modified organisms by synthetic protein design. *Nature, 518*(7537), 55–60. DOI: https://doi.org/10.1038/nature14121

Mann, A. (2016). The power of prediction markets. *Nature News, 538*(7625), 308. DOI: https://doi.org/10.1038/538308a

Marris, C., Langford, I. H., & O'Riordan, T. (1998). A quantitative test of the cultural theory of risk perceptions: Comparison with the psychometric paradigm. *Risk analysis, 18*(5), 635–47.

Martinson, B. C. (2017). Give researchers a lifetime word limit. *Nature, 550*(7676), 303. DOI: https://doi.org/10.1038/550303a

Masood, E. (2016). *The Great Invention: The Story of GDP and the Making and Unmaking of the Modern World.* 1 edition. New York: Pegasus Books.

Mazzucato, M. (2011). *The Entrepreneurial State.* London: Demos.

McGilvray, A. (2014). The limits of excellence. *Nature, 511*(5710), S64–S66.

McNeil Jr, D. G. (2015, August 14). A New Strain of Yeast Can Produce Narcotics. *New York Times.*

McWilliams, T. G., & Suomalainen, A. (2019). Mitochondrial DNA can be inherited from fathers, not just mothers. *Nature, 565*(7739), 296. DOI: https://doi.org/10.1038/d41586-019-00093-1

Merton, R. K. (1942). Science and technology in a democratic order. *Journal of Legal and Political Sociology, 1*, 115–126.

Merton, R. K. (1968). The Matthew Effect in Science: The reward and communication systems of science are considered. *Science, 159*(3810), 56–63. DOI: https://doi.org/10.1126/science.159.3810.56

Mervis, J. (2014). Peering Into Peer Review. *Science, 343*(6171), 596–598.

Meyer, R., & Kunreuther, H. (2017). *The Ostrich Paradox: Why We Underprepare for Disasters.* Wharton Digital Press.

Meyer, S., & Held, L. (2014). Power-law models for infectious disease spread. *The Annals of Applied Statistics, 8*(3), 1612–1639. DOI: https://doi.org/10.1214/14-AOAS743

Miller, S. et al. (2009). Sustaining progress in the life sciences: strategies for managing dual use research of concern--progress at the national level. *Biosecurity and bioterrorism: Biodefense strategy, practice, and science, 7*(1), 93–100. DOI: https://doi.org/10.1089/bsp.2009.1018

Möller, N. (2012). The Concepts of Risk and Safety. In S. Roeser et al. (Eds.), *Handbook of Risk Theory* (p. 1187). Dordrecht: Springer.

Möller, N., & Hansson, S. O. (2008). Principles of engineering safety: Risk and uncertainty reduction. *Reliability Engineering & System Safety, 93*(6), 798–805. DOI: https://doi.org/10.1016/j.ress.2007.03.031

Mollison, D. & Din, S. U. (1993). Deterministic and stochastic models for the seasonal variability of measles transmission. *Mathematical biosciences, 117*(1–2), 155–77.

Moodie, M. (2012). *Managing the Security Risks of Emerging Technologies.*

Mooney, C. (2010). *Do Scientists Understand the Public? American Academy of Arts & Sciences.* Cambridge, MA: American Academy of Arts & Sciences.

Morens, D. M., Subbarao, K., & Taubenberger, J. K. (2012). Engineering H5N1 avian influenza viruses to study human adaptation. *Nature, 486*(7403), 335–40.

Morgan, M. G. (2014). Use (and abuse) of expert elicitation in support of decision making for public policy. *Proceedings of the National Academy of Sciences, 111*(20), 7176–84. DOI: https://doi.org/10.1073/pnas.1319946111

Morgan, M. G. (2015). Our Knowledge of the World is Often Not Simple: Policymakers Should Not Duck that Fact, But Should Deal with It. *Risk analysis, 35*(1), 19–20. DOI: https://doi.org/10.1111/risa.12306

Morgan, M. G., Henrion, M., & Small, M. (1990). *Uncertainty: A Guide to Dealing with Uncertainty in Quantitative Risk and Policy Analysis.* New York: Cambridge University Press.

Moser, C. et al. (2012). The Crucial Role of Nomothetic and Idiographic Conceptions of Time: Interdisciplinary Collaboration in Nuclear Waste Management. *Risk Analysis, 32*(1), 138–154. DOI: https://doi.org/10.1111/j.1539-6924.2011.01639.x

Muir, J. (1911). *My First Summer in the Sierra.* Boston: Houghton Mifflin.

Mulkay, M. (1979). *Science and the Sociology of Knowledge.* London: George Allen & Unwin.

Muller, J. Z. (2018). *The Tyranny of Metrics.* Princeton: Princeton University Press.

Nabel, G. J., & Shiver, J. W. (2019). All for one and one for all to fight flu. *Nature, 565*(7737), 29. DOI: https://doi.org/10.1038/d41586-018-07654-w

Naess, A. (1973). The shallow and the deep, long-range ecology movement. A summary. *Inquiry, 16*(1–4), 95–100. DOI: https://doi.org/10.1080/00201747308601682

Nakayachi, K. (2013). The unintended effects of risk-refuting information on anxiety. *Risk analysis, 33*(1), 80–91. DOI: https://doi.org/10.1111/j.1539-6924.2012.01852.x

National Academies of Sciences, Engineering, and Medicine. (2018a). *Biodefense in the Age of Synthetic Biology.* Washington, DC: National Academies Press. DOI: https://doi.org/10.17226/24890

National Academies of Sciences, Engineering, and Medicine. (2018b). *Governance of Dual Use Research in the Life Sciences: Advancing Global Consensus on Research Oversight: Proceedings of a Workshop.* Washington, DC: National Academies Press. DOI: https://doi.org/10.17226/25154

National Academies of Sciences, Engineering, and Medicine. (2016). *Gene Drives on the Horizon: Advancing Science, Navigating Uncertainty, and Aligning Research with Public Values.* Washington, DC: National Academies Press. DOI: https://doi.org/10.17226/23405

National Academies of Sciences, Engineering, and Medicine. (2017a). *Dual Use Research of Concern in the Life Sciences: Current Issues and Controversies.* Washington, DC: National Academies Press. DOI: https://doi.org/10.17226/24761

National Academies of Sciences, Engineering, and Medicine. (2017b). *Human Genome Editing: Science, Ethics, and Governance.* Washington, DC: The National Academies Press. DOI: https://doi.org/10.17226/24623

National Science Board. (2016). *Science and Engineering Indicators 2016 (NSB-2016-1).* Arlington, VA: National Science Foundation, 897–897.

Newman, M. E. J. (2005). Power laws, Pareto distributions and Zipf's law. *Contemporary Physics, 46*(5), 323–351. DOI: https://doi.org/10.1080/00107510500052444

Nichols, A. L., & Zeckhauser, R. J. (1988). The perils of prudence: How conservative risk assessments distort regulation. *Regulatory toxicology and pharmacology: RTP, 8*(1), 61–75.

Nichols, T. M. (2017). *The Death of Expertise: The Campaign Against Established Knowledge and Why it Matters.* 1 edition. New York, NY: Oxford University Press.

Nightingale, P., & Martin, P. (2004). The myth of the biotech revolution. *Trends in Biotechnology, 22*(11), 564–569. DOI: https://doi.org/10.1016/j.tibtech.2004.09.010

NIH. (2014). *Conducting Risk and Benefit Analysis on Gain-of Function Research Involving Pathogens with Pandemic Potential.* Available at: https://www.fbo.gov/?s=opportunity&mode=form&id=c134018fd1d008c582b7755be1fc1c06&tab=core&_cview=0 (Accessed: 8 April 2015).

Van Noorden, R. (2015). Seven thousand stories capture impact of science. *Nature, 518*(7538), 150–1.

Nordhaus, W. D. (2007). A Review of the Stern Review on the Economics of Climate Change. *Journal of Economic Literature, 45*(3), 686–702.

Normile, D. (2016). Epidemic of Fear. *Science, 351*(6277), 1022–1023.

Norton, M. I., Mochon, D., & Ariely, D. (2012). The IKEA effect: When labor leads to love. *Journal of Consumer Psychology, 22*(3), 453–460.

Noyce, R. S., Lederman, S., & Evans, D. H. (2018). Construction of an infectious horsepox virus vaccine from chemically synthesized DNA fragments. *PLOS ONE, 13*(1), e0188453. DOI: https://doi.org/10.1371/journal.pone.0188453

NRC. (2015). *Potential Risks and Benefits of Gain-of-Function Research: Summary of a Workshop.* Washington, DC: National Academies Press.

NRC. (2016) *Gain-of-Function Research: Summary of the Second Symposium, March 10-11, 2016.* Washington, DC: National Academies Press. DOI: https://doi.org/10.17226/23484

NSABB. (2007). *Proposed Framework for the Oversight of Dual Use Life Sciences Research: Strategies for Minimizing the Potential Misuse of Research Information.* Bethesda, MD: National Science Advisory Board for Biosecurity.

NSABB. (2016). *Recommendations for the Evaluation and Oversight of Proposed Gain-of-Function Research.* Bethesda, MD.

Nussbaum, M. C. (1990). *Love's Knowledge.* New York: Oxford University Press. DOI: https://doi.org/0195054571

Nyborg, K. et al. (2016). Social norms as solutions. *Science, 354*(6308), 42–43. DOI: https://doi.org/10.1126/science.aaf8317

O'Hagan, A., & Oakley, J. E. (2004). Probability is perfect, but we can't elicit it perfectly. *Reliability Engineering & System Safety, 85*(1–3), 239–248. DOI: https://doi.org/10.1016/j.ress.2004.03.014

Orange, R. (2014). Medicine. Pioneering womb transplant trial highlights risks and ethical dilemmas. *Science, 343*(6178), 1418–9. DOI: https://doi.org/10.1126/science.343.6178.1418

Ord, T., Hillerbrand, R., & Sandberg, A. (2010). Probing the improbable: methodological challenges for risks with low probabilities and high stakes. *Journal of Risk Research, 13*(2), 191–205. DOI: https://doi.org/10.1080/13669870903126267

Osterholm, M. T., & Olshaker, M. (2017). *Deadliest Enemy.* New York: Little, Brown, and Company.

OSTP. (2014). *White House Office of Science and Technology Policy. US Government Gain-of-Function Deliberative Process and Research Funding Pause on Selected Gain-of-Function Research Involving Influenza, MERS, and SARS Viruses. October 17, 2014.* Available at: http://www.phe.gov/s3/dualuse/Documents/ (Accessed: 21 October 2014).

OSTP. (2017). *Recommended Policy Guidance for Departmental Development of Review Mechanisms for Potential Pandemic Pathogen Care and Oversight (P3CO).* Washington, DC.

OTA. (1986). *Research Funding as an Investment: Can We Measure the Returns?* Washington, DC: Office of Technology Assessment.

Owens, B. (2013a). Judgement day. *Nature, 502*(7471), 288–290.

Owens, B. (2013b). Long-term research: Slow science. *Nature News, 495*(7441), 300. DOI: https://doi.org/10.1038/495300a

Oye, K. A., Lawson, J. C. H., & Bubela, T. (2015). Regulate 'home-brew' opiates. *Nature, 521*(7552), 281–283. DOI: https://doi.org/10.1038/521281a

Palese, P., & Wang, T. T. (2012). H5N1 influenza viruses: Facts, not fear. *Proceedings of the National Academy of Sciences, 109*(7), 2211–3. DOI: https://doi.org/10.1073/pnas.1121297109

Parthasarathy, S. (2018). Use the patent system to regulate gene editing. *Nature, 562*(7728), 486. DOI: https://doi.org/10.1038/d41586-018-07108-3

Paté-Cornell, E., & Cox, L. A. (2014). Improving risk management: From lame excuses to principled practice. *Risk analysis, 34*(7), 1228–39. DOI: https://doi.org/10.1111/risa.12241

Patterson, A. P. et al. (2013). A Framework for Decisions About Research with HPAI H5N1 Viruses. *Science, 339*(6123), 1036–1037.

Pedroni, N. et al. (2017). A Critical Discussion and Practical Recommendations on Some Issues Relevant to the Nonprobabilistic Treatment of Uncertainty in Engineering Risk Assessment. *Risk Analysis, 37*(7), 1315–1340. DOI: https://doi.org/10.1111/risa.12705

Perkins, D., Danskin, K., & Rowe, A. E. (2017). Fostering an International Culture of Biosafety, Biosecurity, and Responsible Conduct in the Life Sciences. *Science & Diplomacy*, (September). Available at: http://www. sciencediplomacy.org/article/2017/biosafety (Accessed: 23 May 2018).

Phillips, T. et al. (2018). America COMPETES at 5 years: An Analysis of Research-Intensive Universities' RCR Training Plans. *Science and Engineering Ethics, 24*(1), 227–249. DOI: https://doi.org/10.1007/s11948-017-9883-5

Phuc, H. K. et al. (2007). Late-acting dominant lethal genetic systems and mosquito control. *BMC Biology, 5,* 11. DOI: https://doi.org/10.1186/1741-7007-5-11

de Picoli Junior, S. et al. (2011). Spreading Patterns of the Influenza A (H1N1) Pandemic. *PLOS ONE, 6*(3), e17823. DOI: https://doi.org/10.1371/journal.pone.0017823

Pielke Jr, R. A. (2007). *The Honest Broker: Making Sense of Science in Policy and Politics.* Cambridge University Press.

Pilkey, O. H., & Pilkey-Jarvis, L. (2007). *Useless Arithmetic: Why Environmental Scientists Can't Predict the Future.* New York: Columbia University Press.

Pinker, S. (2015, August 1). The moral imperative for bioethics. *The Boston Globe.*

Pinker, S. (2018). *Enlightenment Now: The Case for Reason, Science, Humanism, and Progress.* 1st Edition edition. New York, New York: Viking.

Pohjola, M. V. et al. (2012). State of the art in benefit-risk analysis: Environmental health. *Food and Chemical Toxicology, 50*(1), 40–55. DOI: https://doi.org/10.1016/j.fct.2011.06.004

Pollack, R. (2015). Eugenics lurk in the shadow of CRISPR. *Science, 348*(6237), 871. DOI: https://doi.org/10.1126/science.348.6237.871-a

Pontin, J. (2012). Why we can't solve big problems. *MIT Technology Review, 115*(6).

Poulin, B., Lefebvre, G., & Paz, L. (2010). Red flag for green spray: Adverse trophic effects of Bti on breeding birds. *Journal of Applied Ecology, 47*(4), 884–889. DOI: https://doi.org/10.1111/j.1365-2664.2010.01821.x

Press, W. H. (2013). What's So Special About Science (And How Much Should We Spend on It?). *Science, 342*(6160), 817–822.

Quigley, J., & Revie, M. (2011). Estimating the probability of rare events: Addressing zero failure data. *Risk analysis, 31*(7), 1120–32. DOI: https://doi.org/10.1111/j.1539-6924.2010.01568.x

Quinn, S. C. et al. (2011). Racial disparities in exposure, susceptibility, and access to health care in the US H1N1 influenza pandemic. *American journal of public health, 101*(2), 285–93. DOI: https://doi.org/10.2105/AJPH.2009.188029

Rappert, B. (2014). Why has Not There been More Research of Concern? *Frontiers in public health, 2*(74). DOI: https://doi.org/10.3389/fpubh.2014.00074

Rappert, B., & Selgelid, M. J. (2013). *On the Dual Uses of Science and Ethics: Principles, Practices, and Prospects.* ANU Press (Practical Ethics and Public Policy). DOI: https://doi.org/http://doi.org/10.22459/DUSE.12.2013

Rappuoli, R., Bloom, D. E., & Black, S. (2017). Deploy vaccines to fight superbugs. *Nature, 552*(7684), 165–167. DOI: https://doi.org/10.1038/d41586-017-08323-0

Rask, M. (2013). The tragedy of citizen deliberation – two cases of participatory technology assessment. *Technology Analysis & Strategic Management, 25*(1), 39–55. DOI: https://doi.org/10.1080/09537325.2012.751012

Reaven, S. J. (1987). Science Literacy Needs of Public Involvement Programs. In L. J. Waks (Ed.), *Technological Literacy. Proceedings of the National Science, Technology and Society (STS) Conference* (pp. 347–356). Washington, DC: STS Press.

Reeves, R. G. et al. (2018). Agricultural research, or a new bioweapon system? *Science, 362*(6410), 35–37. DOI: https://doi.org/10.1126/science.aat7664

Refsgaard, J. C. et al. (2006). A framework for dealing with uncertainty due to model structure error. *Advances in Water Resources, 29*(11), 1586–1597. DOI: https://doi.org/10.1016/j.advwatres.2005.11.013

Refsgaard, J. C. et al. (2007). Uncertainty in the environmental modelling process – A framework and guidance. *Environmental Modelling & Software, 22*(11), 1543–1556. DOI: https://doi.org/10.1016/j.envsoft.2007.02.004

Rembold, C. M. (2004), The health benefits of eating salmon. *Science, 305*(5683), 475; author reply 475. DOI: https://doi.org/10.1126/science.305.5683.475b

Renn, O., & Graham, P. (2005). *Risk Governance – Towards an Integrative Approach.* 1. Geneva: International Risk Governance Council.

Renn, O., & Klinke, A. (2004). Systemic risks: A new challenge for risk management. *EMBO reports, 5*(S1), S41-6. DOI: https://doi.org/10.1038/sj.embor.7400227

Resnik, D. B. (2010). Can Scientists Regulate the Publication of Dual Use Research? *Studies in ethics, law, and technology, 4*(1). DOI: https://doi.org/10.2202/1941-6008.1124

Resnik, D. B., Barner, D. D., & Dinse, G. E. (2011). Dual-use review policies of biomedical research journals. *Biosecurity and bioterrorism : biodefense strategy, practice, and science, 9*(1), 49–54. DOI: https://doi.org/10.1089/bsp.2010.0067

Revill, J., & Jefferson, C. (2013). Tacit knowledge and the biological weapons regime. *Science and Public Policy, 41*(5), 597–610. DOI: https://doi.org/10.1093/scipol/sct090

Rey, F., Schwartz, O., & Wain-Hobson, S. (2013). Gain-of-function research: unknown risks. *Science, 342*(6156), 311. DOI: https://doi.org/10.1126/science.342.6156.311-a

Rhodes, C. J., & Anderson, R. M. (1996). Power laws governing epidemics in isolated populations. *Nature, 381*(6583), 600–602. DOI: https://doi.org/10.1038/381600a0

Richardson, S. M. et al. (2017). Design of a synthetic yeast genome. *Science, 355*(6329), 1040–1044. DOI: https://doi.org/10.1126/science.aaf4557

Ridley, M. (2010). *The Rational Optimist: How Prosperity Evolves.* New York: HarperCollins.

Rifkin, J. (1995). *The End of Work: The Decline of the Global Labor Force and the Dawn of the Post-Market Era.* New York: G.P. Putnam's Sons.

Robberstad, B. (2005). QALY vs DALY vs LYs gained: What are the difference, and what difference do they make for health care priority setting? *Norsk Epidemiologi, 15*(2), 183–191.

Roberts, R., & Relman, D. A. (2015). *Letter to NSABB Chair, February 24, 2015.* Available at: http://news.sciencemag.org/sites/default/files/Roberts Relman Letter, as sent to NSABB Chair, February 24, 2015, 14h30 U.S. Eastern.pdf.

Rocca, E., & Andersen, F. (2017). How biological background assumptions influence scientific risk evaluation of stacked genetically modified plants: an analysis of research hypotheses and argumentations. *Life Sciences, Society and Policy, 13*(1), 11. DOI: https://doi.org/10.1186/s40504-017-0057-7

Roeser, S. (2011). Nuclear energy, risk, and emotions. *Philosophy and Technology, 24*(2), 197–201.

Roeser, S., & Pesch, U. (2016). An Emotional Deliberation Approach to Risk. *Science, Technology, & Human Values, 41*(2), 274–297. DOI: https://doi. org/10.1177/0162243915596231

Rooney, P. (1992). On Values in Science: Is the Epistemic/Non-Epistemic Distinction Useful? In *PSA: Proceedings of the Biennial Meeting of the Philosophy of Science Association.* Chicago: University of Chicago Press, 13–22.

Roosth, S. (2017). *Synthetic: How Life Got Made.* 1 edition. Chicago; London: University of Chicago Press.

Rosling, H., Rönnlund, A. R., & Rosling, O. (2018). *Factfulness: Ten Reasons We're Wrong About the World-and Why Things Are Better Than You Think.* New York: Flatiron Books.

Rossi, J. (1997). Participation Run Amok: The costs of mass participation for deliberative agency decisionmaking. *Northwestern University Law Review, 92*(1), 173–249.

Rotman, D. (2013). How Technology Is Destroying Jobs. *MIT Technology Review, 116*(4), 28–35.

Roy, M. et al. (2014). Epidemic cholera spreads like wildfire. *Scientific Reports, 4,* 3710. DOI: https://doi.org/10.1038/srep03710

Rozell, D. J. (2018). The Ethical Foundations of Risk Analysis. *Risk Analysis.* DOI: https://doi.org/10.1111/risa.12971

Rozo, M., & Gronvall, G. K. (2015). The Reemergent 1977 H1N1 Strain and the Gain-of-Function Debate. *mBio, 6*(4). DOI: https://doi.org/10.1128/mBio.01013-15

Rubinstein, A. (2006). Dilemmas of an economic theorist. *Econometrica, 74*(4), 865–883.

Ruckelshaus, W. D. (1984). Risk in a Free Society. *Risk Analysis, 4*(3), 157–162. DOI: https://doi.org/10.1111/j.1539-6924.1984.tb00135.x

Rudner, R. (1953). The Scientist Qua Scientist Makes Value Judgments. *Philosophy of Science, 20*(1), 1–6. DOI: https://doi.org/10.1086/287231

Rudski, J. M. et al. (2011). Would you rather be injured by lightning or a downed power line? Preference for natural hazards. *Judgment and decision making, 6*(4), 314–322.

Russell, C. A. et al. (2014). Improving pandemic influenza risk assessment. *eLife, 3*, e03883. DOI: https://doi.org/10.7554/eLife.03883

Russell, S. (2015). Ethics of artificial intelligence: Take a stand on AI weapons. *Nature, 521*(7553), 415–418. DOI: https://doi.org/10.1038/521415a

Sagan, L. (1967). On the origin of mitosing cells. *Journal of Theoretical Biology, 14*(3), 225-IN6. DOI: https://doi.org/10.1016/0022-5193(67)90079-3

Sahl, J. W. et al. (2016). A Bacillus anthracis Genome Sequence from the Sverdlovsk 1979 Autopsy Specimens. *mBio, 7*(5), e01501-16. DOI: https://doi.org/10.1128/mBio.01501-16

Sale, K. (1995). *Rebels Against the Future.* Reading, MA: Addison-Wesley Publishing.

Sandman, P. M. (2012). *Science versus Spin: How Ron Fouchier and Other Scientists Miscommunicated about the Bioengineered Bird Flu Controversy.* Available at: http://www.psandman.com/articles/Fouchier.htm (Accessed: 2 February 2015).

Sarewitz, D. (2004). How science makes environmental controversies worse. *Environmental Science & Policy.* (Science, Policy, and Politics: Learning from Controversy Over The Skeptical Environmentalist), *7*(5), 385–403. DOI: https://doi.org/10.1016/j.envsci.2004.06.001

Sarewitz, D. (2015). Science can't solve it. *Nature, 522*(7557), 413–414. DOI: https://doi.org/10.1038/522413a

Sarewitz, D. (2016). The pressure to publish pushes down quality. *Nature, 533*(7602), p. 147. DOI: https://doi.org/10.1038/533147a

Sarewitz, D., & **Pielke Jr, R. A.** (2007). The neglected heart of science policy: Reconciling supply of and demand for science. *Environmental Science & Policy, 10*(1), 5–16.

Scharre, P. (2018). *Army of None: Autonomous Weapons and the Future of War.* 1 edition. New York: W. W. Norton & Company.

Scheiner, S. M., & **Bouchie, L. M.** (2013). The predictive power of NSF reviewers and panels. *Frontiers in Ecology and the Environment, 11*(8), 406–407.

Scheufele, D. A. (2006). Messages and heuristics: How audiences form attitudes about emerging technologies. In J. Turney (Ed.), *Engaging science: Thoughts, deeds, analysis and action* (pp. 20–25). London: The Wellcome Trust.

Scheufele, D. A. et al. (2017). U.S. attitudes on human genome editing. *Science, 357*(6351), 553–554. DOI: https://doi.org/10.1126/science.aan3708.

Schoch-Spana, M. (2015). Public Engagement and the Governance of Gain-of-Function Research., *Health Security, 13*(2), 1–5. DOI: https://doi.org/10.1089/hs.2015.0005

Schoch-Spana, M. et al. (2017). Global Catastrophic Biological Risks: Toward a Working Definition. *Health Security, 15*(4), 323–328. DOI: https://doi.org/10.1089/hs.2017.0038

Scotchmer, S. (2004). *Innovation And Incentives*. Cambridge, MA: MIT Press.

Seidl, R. et al. (2013). Perceived risk and benefit of nuclear waste repositories: Four opinion clusters. *Risk Analysis, 33*(6), 1038–48. DOI: https://doi.org/10.1111/j.1539-6924.2012.01897.x

Selgelid, M. J. (2009). Governance of dual-use research: An ethical dilemma. *Bulletin of the World Health Organization, 87*(9), 720–723. DOI: https://doi.org/10.2471/BLT.08.051383

Selgelid, M. J. (2016). Gain-of-Function Research: Ethical Analysis. *Science and Engineering Ethics, 22*(4), 923–964. DOI: https://doi.org/10.1007/s11948-016-9810-1

Sen, A. (1981). *Poverty and Famines: An Essay on Entitlement and Deprivation*. Oxford: Clarendon Press.

Servick, K. (2016). Winged warriors. *Science, 354*(6309), 164–167. DOI: https://doi.org/10.1126/science.354.6309.164

Shafer, G. (1976). *A Mathematical Theory of Evidence*. Princeton, NJ: Princeton University Press.

Shafer, G. (1990). Perspectives on the theory and practice of belief functions. *International Journal of Approximate Reasoning, 4*(5–6), 323–362. DOI: https://doi.org/10.1016/0888-613X(90)90012-Q

Shah, S. K. et al. (2018). Bystander risk, social value, and ethics of human research. *Science, 360*(6385), 158–159. DOI: https://doi.org/10.1126/science.aaq0917

Shechtman, D. et al. (1984). Metallic Phase with Long-Range Orientational Order and No Translational Symmetry. *Physical Review Letters, 53*(20), 1951–1953. DOI: https://doi.org/10.1103/PhysRevLett.53.1951

Shen, C. et al. (2017). A multimechanistic antibody targeting the receptor binding site potently cross-protects against influenza B viruses. *Science Translational Medicine, 9*(412). DOI: https://doi.org/10.1126/scitranslmed.aam5752

Shrader-Frechette, K. S. (1986). The Conceptual Risks of Risk Assessment. *IEEE Technology and Society Magazine, 5*(2), 4–11. DOI: https://doi.org/10.1109/MTAS.1986.5010007

Shrader-Frechette, K. S. (1991). *Risk and Rationality: Philosophical Foundations for Populist Reforms*. Berkeley: University of California Press.

Siler, K., Lee, K., & Bero, L. (2014). Measuring the effectiveness of scientific gatekeeping. *Proceedings of the National Academy of Sciences, 112*(2), p. 201418218.

Singer, P. W. (2009). *Wired for War: The Robotics Revolution and Conflict in the 21st Century*. Reprint edition. New York, NY: Penguin Books.

Singer, P. W., & Friedman, A. (2014). *Cybersecurity and Cyberwar*. Oxford: Oxford University Press.

Sipp, D., & Pei, D. (2016). Bioethics in China: No wild east. *Nature, 534*(7608), 465–467. DOI: https://doi.org/10.1038/534465a

Sjöberg, L. (1998). World Views, Political Attitudes and Risk Perception. *Risk: Health, Safety & Environment, 9,* 137–152.

Sjöberg, L. (2000). Factors in risk perception. *Risk Analysis, 20*(1), 1–11.

Sjöberg, L. (2002). Attitudes toward technology and risk: Going beyond what is immediately given. *Policy Sciences, 35*(4), 379–400. DOI: https://doi.org/10.1023/A:1021354900928

Slovic, P. (1987). Perception of Risk. *Science, 236*(4799), 280–285.

Slovic, P. (2000). *The perception of risk.* London: Earthscan.

Slovic, P. et al. (2004). Risk as analysis and risk as feelings: some thoughts about affect, reason, risk, and rationality. *Risk Analysis, 24*(2), 311–22. DOI: https://doi.org/10.1111/j.0272-4332.2004.00433.x

Slovic, P. et al. (2007). The affect heuristic. *European Journal of Operational Research, 177*(3), 1333–1352. DOI: https://doi.org/10.1016/j.ejor.2005.04.006

Slovic, P., Fischhoff, B., & Lichtenstein, S. (1980). Facts and Fears: Understanding Perceived Risk. In R. C. Schwing & W. A. Albers Jr. (Eds.), *Societal Risk Assessment: How Safe is Safe Enough?* (pp. 181–216). New York: Plenum Press.

Small, M. J., Güvenç, Ü., & DeKay, M. L. (2014). When can scientific studies promote consensus among conflicting stakeholders? *Risk analysis, 34*(11), 1978–94. DOI: https://doi.org/10.1111/risa.12237

Smith, M. R., & Marx, L. (1994). *Does Technology Drive History? The Dilemma of Technological Determinism.* Cambridge, MA: MIT Press.

Smithson, M., & Ben-Haim, Y. (2015). Reasoned Decision Making Without Math? Adaptability and Robustness in Response to Surprise. *Risk analysis, 35*(10), 1911–8. DOI: https://doi.org/10.1111/risa.12397

Sneed, A. (2017). *Mail-Order CRISPR Kits Allow Absolutely Anyone to Hack DNA, Scientific American.* Available at: https://www.scientificamerican.com/article/mail-order-crispr-kits-allow-absolutely-anyone-to-hack-dna/ (Accessed: 10 May 2018).

Solow, R. M. (1957). Technical Change and the Aggregate Production Function. *The Review of Economics and Statistics, 39*(3), 312–320.

Song, G. (2014). Understanding public perceptions of benefits and risks of childhood vaccinations in the United States. *Risk Analysis, 34*(3), 541–555. DOI: https://doi.org/10.1111/risa.12114

Srinivasan, R., & Natarajan, S. (2012). Developments in inherent safety: A review of the progress during 2001–2011 and opportunities ahead. *Process Safety and Environmental Protection, 90*(5), 389–403. DOI: https://doi.org/10.1016/j.psep.2012.06.001

Stern, N. (2006). *Stern Review on the Economics of Climate Change.* Edited by N. N. H. Stern. Cambridge, UK: Cambridge University Press.

Stern, N. & Taylor, C. (2007). Climate change: Risk, ethics, and the Stern Review. *Science, 317*(5835), 203–4. DOI: https://doi.org/10.1126/science.1142920

Suk, J. E. et al. (2011). Dual-use research and technological diffusion: Reconsidering the bioterrorism threat spectrum. *PLoS pathogens, 7*(1), e1001253. DOI: https://doi.org/10.1371/journal.ppat.1001253

Sumathipala, A., Siribaddana, S., & Patel, V. (2004). Under-representation of developing countries in the research literature: Ethical issues arising from a survey of five leading medical journals. *BMC medical ethics, 5*(1), E5. DOI: https://doi.org/10.1186/1472-6939-5-5

Sunstein, C. R. (2005). *Laws of Fear: Beyond the Precautionary Principle.* Cambridge: Cambridge University Press.

Sutton, T. C. et al. (2014). Airborne transmission of highly pathogenic H7N1 influenza virus in ferrets. *Journal of virology, 88*(12), 6623–35. DOI: https://doi.org/10.1128/JVI.02765-13

Taebi, B. (2017). Bridging the Gap between Social Acceptance and Ethical Acceptability. *Risk Analysis, 37*(10), 1817–1827. DOI: https://doi.org/10.1111/risa.12734

Taleb, N. N., Goldstein, D. G., & Spitznagel, M. W. (2009). The six mistakes executives make in risk management. *Harvard Business Review, 87*(10), 78–81.

Tannert, C., Elvers, H.-D., & Jandrig, B. (2007). The ethics of uncertainty. *EMBO Reports, 8*(10), 892–896. DOI: https://doi.org/10.1038/sj.embor.7401072

Teitelbaum, M. S. (2014). *Falling Behind? Boom, Bust and the Global Race for Scientific Talent.* Princeton, NJ: Princeton University Press.

Testa, G., Koon, E. C., & Johannesson, L. (2017). Living Donor Uterus Transplant and Surrogacy: Ethical Analysis According to the Principle of Equipoise. *American Journal of Transplantation, 17*(4), 912–916. DOI: https://doi.org/10.1111/ajt.14086

Tetlock, P. E. (2003). Thinking the unthinkable: Sacred values and taboo cognitions. *Trends in cognitive sciences, 7*(7), 320–324.

Tetlock, P. E., Mellers, B. A., & Scoblic, J. P. (2017). Bringing probability judgments into policy debates via forecasting tournaments. *Science, 355*(6324), 481–483. DOI: https://doi.org/10.1126/science.aal3147

Thicke, M. (2017). Prediction Markets for Science: Is the Cure Worse than the Disease? *Social Epistemology, 31*(5), 451–467. DOI: https://doi.org/10.1080/02691728.2017.1346720

Tierney, J. (2010, May 18). Doomsayers Beware, A Bright Future Beckons. *New York Times.*

Toulmin, S. (1964). The complexity of scientific choice: A stocktaking. *Minerva, 2*(3), 343–359.

Trevan, T. (2015). Biological research: Rethink biosafety. *Nature, 527*(7577), 155–158. DOI: https://doi.org/10.1038/527155a

Tucker, J. B. (2012). *Innovation, Dual Use, and Security: Managing the Risks of Emerging Biological and Chemical Technologies.* Cambridge, MA: MIT Press.

Tucker, W. T., & Ferson, S. (2008). Strategies for risk communication: Evolution, evidence, experience. *Annals of the New York Academy of Sciences, 1128*(631), ix–xii. DOI: https://doi.org/10.1196/annals.1399.000

Tumpey, T. M. et al. (2005). Characterization of the reconstructed 1918 Spanish influenza pandemic virus. *Science, 310*(5745), 77–80. DOI: https://doi.org/10.1126/science.1119392

Tuomisto, J. T. et al. (2004). Risk-benefit analysis of eating farmed salmon. *Science, 305*(5683), 476–7; author reply 476-7. DOI: https://doi.org/10.1126/science.305.5683.476

Tversky, A., & Kahneman, D. (1973). Availability: A heuristic for judging frequency and probability. *Cognitive Psychology, 5*(2), 207–232. DOI: https://doi.org/10.1016/0010-0285(73)90033-9

Tversky, A., & Kahneman, D. (1974). Judgment under Uncertainty: Heuristics and Biases. *Science, 185*(4157), 1124–31. DOI: https://doi.org/10.1126/science.185.4157.1124

Tversky, A., & Kahneman, D. (1981). The framing of decisions and the psychology of choice. *Science, 211*(4481), 453–458.

Tyler, C., & Akerlof, K. (2019). Three secrets of survival in science advice. *Nature, 566*(7743), 175–177. DOI: https://doi.org/10.1038/d41586-019-00518-x

Uhlenhaut, C., Burger, R., & Schaade, L. (2013). Protecting society. *EMBO reports, 14*(1), 25–30. DOI: https://doi.org/10.1038/embor.2012.195

Unwin, S. D. (1986). A Fuzzy Set Theoretic Foundation for Vagueness in Uncertainty Analysis. *Risk Analysis, 6*(1), 27–34. DOI: https://doi.org/10.1111/j.1539-6924.1986.tb00191.x

Usher, S. (2013). *Letters of Note: Correspondence Deserving of a Wider Audience.* Edinburgh: Cannongate.

Vesely, W. E. et al. (1981). *Fault Tree Handbook.* Washington, DC: Nuclear Regulatory Commission.

Vezér, M. et al. (2018). Epistemic and ethical trade-offs in decision analytical modelling. *Climatic Change, 147*(1–2), 1–10. DOI: https://doi.org/10.1007/s10584-017-2123-9

Viscusi, W. K., & Hakes, J. K. (1998). Synthetic risks, risk potency, and carcinogen regulation. *Journal of policy analysis and management, 17*(1), 52–73. DOI: https://doi.org/10.1002/(SICI)1520-6688(199824)17:1<52::AID-PAM4>3.0.CO;2-G

Visser't Hooft, H. P. (1999). *Justice to Future Generations and the Environment.* Dordrecht: Kluwer Academic Publishers.

Vogel, K. M. (2013). Intelligent assessment: Putting emerging biotechnology threats in context. *Bulletin of the Atomic Scientists, 69*(1), 43–52. DOI: https://doi.org/10.1177/0096340212470813

Walker, W. E. et al. (2003). Defining Uncertainty: A Conceptual Basis for Uncertainty Management in Model-Based Decision Support. *Integrated Assessment, 4*(1), 5–17.

Walley, P. (1991). *Statistical Reasoning with Imprecise Probabilities.* London: Chapman and Hall.

Wang, D. et al. (2014). Science communication. Response to Comment on 'Quantifying long-term scientific impact.' *Science, 345*(6193), 149.

Wang, D., Song, C., & Barabási, A.-L. (2013). Quantifying long-term scientific impact. *Science, 342*(6154), 127–32.

Wang, J., Mei, Y., & Hicks, D. (2014). Science communication. Comment on 'Quantifying long-term scientific impact.' *Science, 345*(6193), 149.

Wang, J., Veugelers, R., & Stephan, P. (2016). *Bias against Novelty in Science: A Cautionary Tale for Users of Bibliometric Indicators.* Working Paper 22180. National Bureau of Economic Research. DOI: https://doi.org/10.3386/w22180

Wang, T. T., Parides, M. K., & Palese, P. (2012). Seroevidence for H5N1 Influenza Infections in Humans: Meta-Analysis. *Science, 335*(6075), 1463. DOI: https://doi.org/10.1126/science.1218888

Wang, Z. (2008). *In Sputnik's Shadow: The President's Science Advisory Committee and Cold War America.* Piscataway, NJ: Rutgers University Press.

Watanabe, T. et al. (2014). Circulating Avian Influenza Viruses Closely Related to the 1918 Virus Have Pandemic Potential. *Cell Host & Microbe, 15*(6), 692–705. DOI: http://dx.doi.org/10.1016/j.chom.2014.05.006

Webster, R. G. et al. (1992). Evolution and ecology of influenza A viruses. *Microbiological Reviews, 56*(1), 152–79.

Weichselberger, K. (2000). The theory of interval-probability as a unifying concept for uncertainty. *International Journal of Approximate Reasoning, 24*(2–3), 149–170. DOI: https://doi.org/10.1016/S0888-613X(00)00032-3

Weiman, C. (2007). The 'Curse of Knowledge' or Why Intuition about Teaching Often Fails. *American Physical Society News, 16*(10), 8.

Weinberg, A. M. (1972). Science and trans-science. *Minerva, 10*(2), 209–222. DOI: https://doi.org/10.1007/BF01682418

Weinberg, B. A. et al. (2014). Science funding and short-term economic activity. *Science, 344*(6179), 41–3.

Werner, R. (2015). The focus on bibliometrics makes papers less useful. *Nature, 517*(7534), 245.

Wilson, R., & Crouch, E. A. C. (2001). *Risk-Benefit Analysis.* 2nd edn. Cambridge, MA: Harvard Univsersity Press.

Wilson, R. S., Zwickle, A., & Walpole, H. (2019). Developing a Broadly Applicable Measure of Risk Perception. *Risk Analysis, 39*(4), 777–791. DOI: https://doi.org/10.1111/risa.13207

Wimmer, E. (2006). The test-tube synthesis of a chemical called poliovirus. The simple synthesis of a virus has far-reaching societal implications. *EMBO reports,* 7 Spec No, S3-9. DOI: https://doi.org/10.1038/sj.embor.7400728

Windbichler, N. et al. (2007). Homing endonuclease mediated gene targeting in Anopheles gambiae cells and embryos. *Nucleic Acids Research, 35*(17), 5922–5933. DOI: https://doi.org/10.1093/nar/gkm632

Winkler, R. L. (2015). Equal versus differential weighting in combining forecasts. *Risk Analysis, 35*(1), 16–8. DOI: https://doi.org/10.1111/risa.12302

Winkler, R. L., Smith, J. E., & Fryback, D. G. (2002). The Role of Informative Priors in Zero-Numerator Problems: Being Conservative versus Being Candid. *The American Statistician, 56*(1), 1–4.

Winner, L. (1986). *The Whale and the Reactor: A Search for Limits in an Age of High Technology.* University of Chicago Press.

von Winterfeldt, D., & Edwards, W. (2007). Defining a Decision Analytic Structure. In W. Edwards, R. F. Miles, and D. von Wintereldt (Eds.), *Advances in Decision Analysis* (pp.489–513). New York: Cambridge University Press.

Wright, S. (2006). Terrorists and biological weapons. Forging the linkage in the Clinton Administration. *Politics and the Life Sciences, 25*(1–2), 57–115. DOI: https://doi.org/10.2990/1471-5457(2006)25[57:TABW]2.0.CO;2

Wyatt, E. (2011, February 21). U.S. Sets 21st-Century Goal: Building a Better Patent Office. *New York Times.*

Xie, X.-F. et al. (2011). The role of emotions in risk communication. *Risk Analysis, 31*(3), 450–65. DOI: https://doi.org/10.1111/j.1539-6924.2010.01530.x

Yamamoto, Y. T. (2012). Values, objectivity and credibility of scientists in a contentious natural resource debate. *Public Understanding of Science, 21*(1), 101–25.

Yaqub, O. (2018). Serendipity: Towards a taxonomy and a theory. *Research Policy, 47*(1), 169–179. DOI: https://doi.org/10.1016/j.respol.2017.10.007

Yong, E. (2017, July 11). One Man's Plan to Make Sure Gene Editing Doesn't Go Haywire. *The Atlantic.* Retrieved from https://www.theatlantic.com/science/archive/2017/07/a-scientists-plan-to-protect-the-world-by-changing-how-science-is-done/532962/

Young, A., & Penzenstadler, N. (2015). *Inside America's secretive biolabs, USA Today.* Retrieved from http://www.usatoday.com/story/news/2015/05/28/biolabs-pathogens-location-incidents/26587505/

Young, N. S., Ioannidis, J. P. A., & Al-Ubaydli, O. (2008). Why current publication practices may distort science. *PLoS medicine, 5*(10), e201. DOI: https://doi.org/10.1371/journal.pmed.0050201

Zadeh, L. A. (1965). Fuzzy sets. *Information and Control, 8*(3), 338–353. DOI: https://doi.org/10.1016/S0019-9958(65)90241-X

Zhang, J. et al. (2017). Live birth derived from oocyte spindle transfer to prevent mitochondrial disease. *Reproductive BioMedicine Online, 34*(4), 361–368. DOI: https://doi.org/10.1016/j.rbmo.2017.01.013

Zhang, S. (2018, February 20). A Biohacker Regrets Publicly Injecting Himself With CRISPR. *The Atlantic.* Retrieved from https://www.theatlantic.com/science/archive/2018/02/biohacking-stunts-crispr/553511/

Zolas, N. et al. (2015). Wrapping it up in a person: Examining employment and earnings outcomes for Ph.D. recipients. *Science, 350*(6266), 1367–1371.

Index

A

ability-to-pay measure   34
absolute vs. relative risk   34
aleatory vs. epistemic
    uncertainty   45
anthrax   4, 79, 90, 92
applied vs. basic research
  and academic publications and
      impact   16
  and econometric valuation   12, 25
  public attitudes and values   26
Asilomar model of science
    policy   87
atomic bomb. *see* nuclear weapons
autism   21
autonomous weapons   75
avian influenza
  Fouchier and Kawaoka research
      studies and response   3, 4,
      54, 55, 85, 92
  H5N1   2, 3, 25, 34, 55, 76, 78, 79,
      80, 85, 92, 94

H7N1   3
H7N9   3

B

banned research. *see also* dangerous
    science
  autonomous weapons
      (proposed)   75
  bioweapons   74
  moratoria   7, 71, 82
  reasons for banning   66, 80
basic research
  societal benefits   17
  vs. applied research. *see* applied vs.
      basic research
Bayesian models   48
benefits assessment. *see also* cost-
    benefit analysis
  academic publications and
      impact   14, 16, 17, 19
  appropriate to research types
      25, 26

econometric valuation
methods   13, 19, 26
expert opinion   19, 24
Innovation Union Scoreboard   17
knowledge output, time-lag
effect   16, *see also* 'sleeping
beauty' academic publications
methods, overview   19
multiple metrics methods   18
patent activity   15, 16, 17, 19
priority realignment process   20
public attitudes and values   26
public vs. private benefit of
research   12
research as a jobs program   10,
19, 26
societal benefits. *see* societal
benefits of research
STAR METRICS program   10, 17
value-of-information (VOI)
analysis   21, 53
bias
in expert opinion   22
in risk assessment   32, 52, 60
publication bias(es)   42
Big Data and data mining   39
bioethics   62, 67, *see also* ethics
human genome editing   70, 75
Nuremberg Code. *see* Nuremberg
Code
synthetic biology   72, 83
training dearth   x, 91
womb transplants   69
biosafety
increasing risk   78, 82
international agreements   82
laboratory accidents and errors   4,
54, 55, 79, 80, 90, 92
laboratory closures   4
research guidelines and policies
3, 4, 81, 87, 91, 92
risk management strategies   79,
90, 93, *see also* risk
management

biosecurity and bioterrorism
guidelines and policies   75, 91
risk assessment   76
risk management   79
training dearth   91
bioweapons   74
bird flu. *see* avian influenza , *see* avian
influenza
'blue skies' research. *see* basic
research
bounding analysis   48

C

Cartagena Protocol on Biosafety   82
case studies approach to benefits
assessment   11
CDC (US Centers for Disease
Control and Prevention)   4,
80, 90, 91
Chinese Academy of Sciences   18
citation metrics   14
climate change research and
policy   15
geoengineering   101
International Panel on Climate
Change   87
cloning   71
complexity
analytical methods   40
in risk assessment   41
conflict of interest, in expert
opinion   22
Convention on Biological
Diversity   82
Cooper, Marty   104
cost-benefit analysis   19
Craven, T.A.M.   104
CRISPR   71, 97
CRISPR-Cas9   72, 75, 78, 83,
96, 97
cultural cognition theory   62
cultural theory of risk   60, 61
cybersecurity   100

D

dangerous science   x, xi
  and emerging technologies   75
  management. *see* risk management
  predictive power   100
  principle   68
data
  incompleteness problem   39
  rare events   44
  risk assessment   44, 52, 53
data mining   39
deterministic risk analysis   44
discount rate, in risk assessment   35
Doudna, Jennifer   83
dual-use research   3, 4, 21, 77

E

Ebola   91
econometric valuation methods   13,
    19, 26
economic efficiency   37
economic rate of return for
    research   13
egalitarians, risk attitudes   60
Einsten, Albert   104
empirical vs. theoretical
    models   40
empiricism   39
EPA (US Environmental Protection
    Agency)   53
epidemics   38, *see also* pandemics
epistemic vs. aleatory
    uncertainty   45
epistemic vs. non-epistemic
    values   31, 32
epistemology   30
error treatment, in risk
    assessment   38
Esvelt, Keven   98
ethics   105
  and banned research   66
  and risk assessment and

communication   33, 34, 35,
    37, 42, 51
  and technological decisions   57,
    67, 68
  and value judgements   31
  bioethics. *see* bioethics
  training dearth   x, 91
evolution of technology   67, 75, 76
expectation value. *see* risk assessment
expert opinion
  bias in   22
  conflict of interest   22
  in benefits assessment   19, 24
  in risk assessment   43, 44, 94
  science policy analyses (US)   22

F

fatalists, risk attitudes   60
FDA (US Food and Drug
    Administration)   4, 35, 49
Feyerabend, Paul   42
Fischhoff, Baruch   7
Forest, Lee de   104
Fouchier, Ron   2, 3, 54, 84
Frieden, Thomas   4
Fukushima nuclear accident, cancer
    risk communication   33
funding
  research. *see* research funding
  risk assessments

G

gain-of-function research
  see also biosafety; biosecurity and
    bioterrorism
  alternatives   92
  biosafety measures   82
  definition   3
  moratoria   7, 71, 82
  ongoing challenges   8
  risk-benefit analysis   4, 6, 25, 55, 85
Gates, Bill   104
  Gates Foundation   82

GDP (gross domestic product)   11,
        12, 14
genetic engineering and genome
        editing
    'DIY', 118.170   79
    E. coli bacteria   78
    gene drives   98
    humans   72, 75, 87
    international agreements   82
    patents as regulation tool   79
    risk assessment   95
    synthetic biology   73, 78, 82

H

heuristic-to-systematic thinking
        model   58
HHS (US department of Health and
        Human Services)   3, 7, 92
hierarchists, risk attitudes   60
horsepox   7
human cloning   71
human genome editing   72, 75, 87

I

individualists, risk attitudes   60
influence diagrams   19
influenza
    accidental exposure   54, 80
    avian influenza. *see* avian influenza
    biosafety risk analysis   5
    pandemics   2, 3, 25, 54, 55, 91
    vaccines   25, 92
information. *see* intellectual property;
        publication(s); VOI (value-of-
        information) analysis
inherent safety approach to risk
        management   91, 93, 95, 98
Innovation Union Scoreboard   17
intellectual property
    patents. *see* patent activity
    trade secrets   15
IVF (*in vitro* fertilization),
        three-person   69, 70

J

Jiankui, He   71
jobs program, research as   10,
        19, 26
Jobs, Steve   104
Jonas, Hans   88

K

Kawaoka, Yoshihiro   1, 3, 92

L

liability insurance   86
literature, and public
        engagement   88
lobotomies   65
low-probability events. *see* rare
        events
Luddite rebellion   59
lumped uncertainty   47

M

macroeconomic approach to benefits
        assessment   12
managing dangerous science. *see* risk
        management
media coverage of avian influenza
        studies   1
meta-analyses   42
Metcalfe, Robert   104
microeconomic approach to benefits
        assessment   11
modernism   58
Monte Carlo models   48
morality. *see* ethics; values
moratoria   7, 71, 82, *see also* banned
        research
mosquitoes   96, 97
Musk, Elon   105

N

NASA (US National Aeronautics and
        Space Administration)   10, 27

National Institute on Drug
     Abuse 73
negative value of information
     (VOI) 21, see also VOI
     (value-of-information)
     analysis
NIH (National Institute for Health)
  biosafety research guidance 3,
     54, 81
  documentary analysis 3
  gain-of-function research
     moratorium 6
  peer review of research
     proposals 23
  research funding 73, 82
NRC (National Research Council)
  risk-benefit analysis 6, 85
  symposium 54, 76
NRC (US Nuclear Regulatory
     Commission) 101
NSABB (US National Science
     Advisory Board for
     Biosecurity)
  publication restriction
     recommendation 2
  risk-benefit analysis 6, 85, 94
NSF (National Science Foundation)
  peer review of research
     proposals 23
  research funding 73
nuclear power industry
  Albert Einstein on 104
  Fukushima nuclear accident 33
  risk assessment 45, 101
  risk management 89
nuclear weapons 66
Nuremberg Code 92

O

Occam's razor 40
Office of Science and Technology
     Policy, US 6
opioid bioengineering 73

P

pandemics see also epidemics
  influenza 2, 3, 25, 54, 55, 91
  potentially pandemic
     pathogens 4, 81, 92
  risk assessment 38, 94
Pascal, Blaise xi
patent activity
  as research assessment metric 15,
     16, 17, 19
  fluctuations 15
peer review
  in Chinese Academy of
     Sciences 18
  in NIH 23
  in NSF 23
  of journal publications 24
polio 102
political ideology, and technological
     risk attitudes 63
politics, rejection of (technocratic
     solutionism) 62
postmodernism 59
prediction markets 86
priority realignment process 20
probabilistic risk analysis 37,
     45, 48
psychometric measures, and risk
     attitudes 61, see also risk-as-
     feeling framework
publication bias(es) 42
publication(s)
  as research assessment metric 14,
     16, 17, 19
  citation metrics 14
  discredited/retracted 21
  increasing risk 100
  journal requirements 102
  peer review process 24
  restricted 3, 99
  volume of 43
public funding of research
  importance 9

National Institute on Drug
    Abuse 73
NIH 73, 82
NSF 73
size (US) 27
public vs. private benefit of
    research 12

R

rare events
  data 44
  risk assessment 45
rational choice theory 36, 60
regulation
  banned research and moratoria.
      see banned research
  genetic engineering and genome
      editing 79
  geographic differences 82
  self-regulation 79, 84, 87, 99, 102
relative vs. absolute risk 34
research funding
  and benefits assessment 27
  and researcher attachment to
      research 99
  and societal benefits of
      research 27
  philanthropic organizations 82
  public funding. *see public* funding
      of research
research proposals
  peer review 23
  risk-benefit analysis 3
risk-as-feeling framework 49, 60, 65
risk assessment
  as risk exploration vs. decision
      tool 94, 98
  boundary (temporal, spatial,
      population) definition 33,
      35, 52, 53
  data collection 44, 52, 53
  deterministic risk analysis 44
  discount rate 35
  error treatment 38

examples 55
improvements
    (recommendations) 95
interpretation 58
level of complexity selection 41
liability insurance 86
model selection 41, 46, 52, 53, 95
multiple methods in 95
prediction markets 86
probabilistic risk analysis 37, 45, 48
public engagement 88, 95, 98
public perceptions and
    response 50, 51, 81, 83
qualitative vs. quantitative
    approaches 36, 54
risk comparisons 51, 52, 53
'science court' model
    (proposed) 80
topic selection and bias 32, 52, 60
uncertainty treatment and
    representation 49, 52, 80, 95
unit of assessment selection 34
utility of 94
value judgements in (data
    collection) 41, 42, 44
value judgements in
    (importance) 57
value judgements in (method/
    model selection) 38, 39, 41
value judgements in
    (overview) 30, 32, 52
value judgements in (risk
    comparisons) 49, 51
value judgements in (uncertainty
    handling) 44
value of life quantification 35
risk attitudes/perceptions
  public perceptions and response to
      risk assessments 50, 51,
      81, 83
  public perceptions and response to
      risk communication 51, 102
  technological risk attitudes. *see*
      technological risk attitudes
  theories 36, 60, 61

risk-benefit analyses. *see also* benefits
	assessment; risk assessment
as risk management strategy   79
challenges   9, 57
gain-of-function research   4, 6, 25,
	55, 85
improvements
	(recommendations) 95
informal examples (quotes)   105
relative vs. absolute risk   34
research proposals   3
risk communication
contextual influence (framing
	effects) 34
public perceptions and
	response 51, 102
relative vs. absolute risk   34
'science court' model
	(proposed) 80
uncertainty treatment   46
risk, definition and
	conceptualization 36
risk management   82
in gene drives   97
inherent safety approach   91, 93,
	95, 98
post hoc management   68, 80, 100
pragmatic approach   80
precautionary approach
	(technology bans) 80
primary and secondary
	prevention 89
public engagement   83, 85, 88,
	90, 98
resiliency approach   88
self-regulation   79, 84, 87, 99, 102
utility maximization (risk-benefit
	analysis) 79
Russell, Stuart   75

S

salmon consumption, risk
	assessment 54
Savulescu, Julian   69

science fiction literature,
	technological optimism and
	skepticism in 64
scientific literacy training   83
self-regulation   79, 84, 87,
	99, 102
sensitivity analysis   47
smallpox   4, 7
societal benefits of research, *see also*
	benefits assessment
and research funding   27
research as a jobs program   10,
	19, 26
time-lag effect   16, *see also*
	'sleeping beauty' academic
	publications
Solow residual   12
space exploration
benefits assessment   25
'space race,' policy analyses   22
STAR METRICS program   10, 17
synthetic biology   73, 78, 82
Synthetic Yeast Genome
	Project 72

T

technological (risk) assessment. *see*
	risk assessment
technological determinism   66, 75
technological evolution and
	inevitability 67, 75, 76
technological optimism
and biotechnology   72, 75, 76
and technological
	inevitability 67
and theoretical frameworks   60,
	62, 63
bias towards   75
definition   59
trends and variability   64, 65
technological revolution   75
technological risk attitudes *see also*
	risk attitudes/perceptions
and political ideology   63

and risk management
    strategies   76
flexibility and changes   65
technological optimism. *see*
    technological optimism
technological skepticism. *see*
    technological skepticism
variation, theories and models   63
technological skepticism
  and biotechnology   72, 75, 76
  and technological inevitability   67
  and theoretical frameworks   60,
        62, 63
  bias against   75, 76
  definition   58, 59
  trends and variability   64, 65
terrorism
  and autonomous weapons   74
  and inherent safety   89
  and negative value of information
        (VOI)   21
  bioterrorism. *see* biosecurity and
        bioterrorism
theoretical vs. empirical
    models   40
time-lag effect of research benefit/
    impact   16, 17, *see also*
    'sleeping beauty' academic
    publications
trade secrets   15
Type I and Type II errors   38

U

UMETRICS program   10
uncertainty
  aleatory vs. epistemic   45
  bounding analysis   48
  in risk assessment   49, 52, 80, 95
  lumped uncertainty   47
  model averaging   48
  non-probabilistic methods   46
  second-order   45

sensitivity analysis   47
uterine transplants   69

V

vaccines   6, 7, 25, 55
  MMR, erroneous
      information   21
  risk-benefits perceptions   60
value of life quantification   35
values
  and ethics   31
  and risk assessment. *see* under risk
      assessment
  definition   30
  epistemic vs. non-epistemic   31, 32
VOI (value-of-information)
      analysis   21, 53
  negative VOI   21

W

weapons
  autonomous weapons   75
  bioweapons   74
  nuclear weapons   66
WHO (World Health Organization)
  cancer risk communication   33
  malaria mortality estimates   96
  publication guidelines   2
willingness-to-pay measure   35
womb transplants   69
Wright, Orville   104

Y

yeast
  opioid bioengineering   73
  Synthetic Yeast Genome
      Project   72

Z

Zuckerberg, Mark   105

www.ingramcontent.com/pod-product-compliance
Lightning Source LLC
Chambersburg PA
CBHW040143270326
41928CB00023B/3335